Klaus H. Sames (ed.)

Applied Cryobiology

Human Biostasis

APPLIED HUMAN CRYOBIOLOGY

Edited by Prof. Dr. Klaus H. Sames

ISSN 2195-5700

1   *Klaus H. Sames (ed.)*
       Applied Cryobiology – Human Biostasis
       ISBN 978-3-8382-0458-1

Klaus H. Sames (ed.)

APPLIED CRYOBIOLOGY
Human Biostasis

ibidem-Verlag
Stuttgart

Bibliografische Information der Deutschen Nationalbibliothek
Die Deutsche Nationalbibliothek verzeichnet diese Publikation in der Deutschen Nationalbibliografie; detaillierte bibliografische Daten sind im Internet über http://dnb.d-nb.de abrufbar.

Bibliographic information published by the Deutsche Nationalbibliothek
Die Deutsche Nationalbibliothek lists this publication in the Deutsche Nationalbibliografie; detailed bibliographic data are available in the Internet at http://dnb.d-nb.de.

Cover illustrations:
(1) Logo Spirale © sgm. Werbeagentur / Fotolia.com
(2) Introduction of increasing glycerol concentrations in the perfusion of a cryonics patient. Schematic sketch hand drawn by Ben Best (http://www.benbest.com/cryonics/protocol.html#CPA_perfusion).
The upper hose comes from the femoral vein; its upper branch leads into a wast container. This way part of the fluid from the body is disposed as waste, but part is recycled into the stirred reservoir by the lower arm of the hose. So it can mix with pure glycerol from the container at the left. Thus, the discarded fluid is continuosly replaced by glycerol. By this mechanism the concentration of glycerol entering the patient is gradually increased -- what we call "ramped cryoprotectant concentration perfusion".
The pump driving the circuit is not given in this scheme for didactic reasons.
Meanwhile newer cryoprotectants have replaced glycerol because they provide better vitrification concentrating perfusion on the brain.
A closed-circuit perfusion, as illustrated in the diagram, can be set up at low cost for gradual introduction of cryoprotectant into cryonics patients. As shown in the diagram, the perfusion circuit bypasses the heart. Perfusate enters the patient through a cannula in the femoral (leg) artery and exits from a cannula in the femoral vein on the same leg. Flowing upwards (opposite from the usual direction) from the femoral artery and up through the descending aorta, the perfusate enters the arch of the aorta (where blood normally exits the heart), but is blocked from entering the heart. Instead, the perfusate flows (in the usual direction) through the distribution arteries of the aorta, to the upper body, including the head and brain. Returning in the veins (in the usual direction), the perfusate nonetheless again bypasses the heart and flows downward (opposite from the usual direction) to the femoral vein where it exits. Note that the lower body will not be well perfused, and the leg with the femoral cut-downs will not be perfused. The stirred reservior will initially have only carrier solution, and no cryoprotectant.
(3) Background: body tanks at Cryonics Institute (archive of the editor).

∞

Gedruckt auf alterungsbeständigem, säurefreien Papier
Printed on acid-free paper

ISSN: 2195-5700

ISBN-13: 978-3-8382-0458-1

© *ibidem*-Verlag
Stuttgart 2013

Alle Rechte vorbehalten

Printed in Germany

In memoriam Robert Ettinger

Foreword

I have seen recent statements to the effect that approximately nothing has been accomplished in the last 50 years in cryonics or cryobiology. One such said that no organ has survived cryopreservation. In fact, many have, some long ago, as the rat parathyroid and the rat uterus–small, but not microscopic. There are stacks of evidence that our methods have improved and show some efficacy.

Then there are the uploaders whose argument is that the brain is a computer, and any computer can emulate any other. I have discussed this at length in YOUNIVERSE, and produced what seem to me to be irrefutable evidence that uploading (a person or persona (into a digital computer) probably never will be possible, and I say "probably" merely because history teaches caution.

I also remain convinced, even though I can't cite statistics, that a majority of "sensible" people tend to denigrate cryonics because they lump it in with the foolishness of uploading and the like. If someone believed in the Flat Earth, would you tend to value his other opinions?

At one of my Johnny Carson TV appearances in the mid Sixties, one of the guests said, "Don't be surprised if your idea takes 20 years to catch on." I scoffed. How could such crystal clear logic and glorious potential take so long to build a band wagon? And why do bona fide cryonicists such as Saul Kent say that selling cryonics is trying to sell something you don't have to someone who doesn't want it?

My early misguided enthusiasm for rapid growth overlooked a basic fact of psychology, summed in two words, *cultural inertia* (or *cultural legacy*). People generally believe what they have been taught to believe, or what makes them comfortable. They hate responsibility. Dostoyevsky said, "Men prefer peace, even death, to freedom of choice in the knowledge of good and evil." Cryonics carries the burden of total personal responsibility, as well as the potential of undermining (betraying) existing loyalties.

Among those few who try to see cryonics objectively, many have said it's just speculation, not science. Where is the evidence? What have cryonicists contributed to low temperature biology literature?

There was plenty of evidence in the early literature, but for a long time experimental cryonicists offered little. Led by Jerry Leaf and Mike Darwin, Alcor did hypothermic work with dogs, which was morally reprehensible and of little practical value. They did, however, demonstrate that their procedures, using glycerol as cryoprotectant, yielded results better than straight freezing.

CI (Cryonics Institute) for a long time was very small and with few financial resources, so was unable to carry out novel procedures. Eventually this picture improved and we hired professionals to evaluate the results of our procedures. One of those hired was Dr. Yuri Pichugin, then working in the Ukraine in the mid nineties at the world's largest low temerature research institute for cryobiology and cryomedicine. Using the heads of freshly slaughtered sheep, prepared by the methods we were then using for human patients, he and colleagues demonstrated that our results were better than with straight freezing.

There was also some work with rabbits (anaesthetized and never allowed to suffer), and some pieces of brain, after thawing from liquid nitrogen, showed coordinated electrical activity among networks of neurons–an important and impressive result.

Dr. Pichugin's main work for CI was development of a vitrification solution, allowing tissue to be "glassified" rather than crystallized, thus eliminating or reducing ice damage. This work was done mainly with rat brain slices, and showed survival not only by structure but also by function, as measured by the K/Na (potassium/sodium) ratio. A precursor to this work was published in *Cryo-*

biology, the journal of the Society for Cryobiology, with Pichugin as lead author.

Since Pichugin left CI, research continues, some of it by Aschwin and Chana de Wolf, partly subsidized by CI and IS (Immortalist Society). This is further enouragement.

Some of the encouraging recent developments have been renewed and extended interest in my first two books, *The Prospect of Immortality* (Doubleday 1964 et seq) and *Man into Superman* (St.Martin's Press 1972 et seq). Recent editions include Russian, Chinese, and Korean languages. Response will be interesting and possibly important.

Especially encouraging has been the relativity rapid rise of the Russian organization, KrioRus, with a building outside Moscow, Established in 2005, it already (April 2011) has 16 human paients, plus some dogs, cats, and birds. By comparison, it took ten years for CI to get its first two patients. (CI now has over 100 human patients, about the same as Alcor.)

Although both *The Prospect of Immortality* and *Man into Superman* were moderately successful by publishing standards, I was ludicrously wrong in my hopes for their recruitment potential. The second book was written in hope of enticing people into cryonics by visions of a magnificent potential future of boundless possibilities and radical change, including changes in the individual psyche. But guess what–most people don't want radical change. They want the present, gold plated and chocolate covered.

The prospect of radical change is for most something to be feared and detested, not coveted and worth working for. This is really just another variation on the cultural inertia phenomenon.

But we are not playing a dirge. Progress has been dismayingly slow, if you are easily dismayed. But the fact is that, despite the snail's pace, progress has been made and there are signs of acceleration. To get CI's first 36 patients took about 24 years, but the next 67 patients only required about an additional ten years–about 4.5

times faster growth. I call that progress–not the explosive progress some still hope for, but solid progress all the same. How do we explain the improvement?

Human stupidity is formidable, but not invincible. Logic doesn't win all battles, but it helps win some. On our side are logic and lust for life, the latter being overrated but still sometimes felt. Love is also a major factor. Mothers are especially well represented in our patient population.

Also on our side are the quality of our people. Cryonicists come in all shapes and sizes, but the typical member is better educated and more intelligent than the average person. I sometimes say, "Doctors choose cryonics, nine to one." That doesn't mean that 90% of physicians support cryonics, but that a doctor chosen at random is 9 times more likely to advocate cryonics than is a random citizen.

Google "Scientists' open letter on cryonics." Here you will find, over a substantial number of solid signatures, some of them very well known, a sober evaluation of cryonics as a respectable endeavor. If you are looking for credentials, here they are. Don't miss it! We have high hopes for German scientists in cryonics. Very best wishes to you,

June 18th 2011
Robert C.W. Ettinger

Preface

This volume compiles the contributions of a scientific cryonics symposium in Goslar Germany 2010.

In addition scientific papers, reviews and reports on special topics have been included to cover a wider variety of science in cryonics and stimulate research in the fields concerned.

The leading themes are discussion of cryonics, its importance, its biology, its procedures and its progress as related to research in different areas of science and problem solving. These occupy the main corpus of the volume.

The biology of aging and life span extension demonstrates the indispensability of cryonics as a time bridging technology allowing to wait for the long lasting development of life span extension – and rejuvenation methods.

Ethical problems and interactions with familiar philosophies like transhumanism must be respected and allow welcome chances of cooperation.

Contacts to heart surgery and brain surgery allow to profit from disciplines, able to control the success of perfusion methods by maintenance of a patients life.

Transportation of cryonics patients from overseas countries is an actual problem needing new technical solutions.

School teaching may be very important for the acceptance of cryonics. One may assume that, the arrangement with death takes place during childhood and puberty and remains untouched thereafter. Before the advent of cryonics no way out of dying had been imaginable, the result could only be resignation, religiousness or repression of thoughts about ones own death. Cryonics provides the first chance to change this situation and show the adolescent a real way out. Therefore it is urgent to teach cryonics at school.

In this volume we try to do a first step into the many different areas of cryonics research and hope it will stimulate science in this exciting project.

February 2013

Klaus H. Sames

Contents

I   Cryonics

Physical And Biological Aspects Of Renal Vitrification

Gregory M. Fahy, Brian Wowk, Roberto Pagotan, Alice Chang,
John Phan, Bruce Thomson, and Laura Phan*

§1. Introduction

The long-term banking of human organs or their engineered substitutes[1] for subsequent transplantation is a long-sought[2-4] and important[1,2,5-11] goal. Given that the full demand for vital and non-vital organ replacements may be over one million per year in the United States alone, supply chain management issues may become more and more critical as the success of laboratory construct creation increases.[1] Contemplating the possible development of emergency organ replacements with generic allografts without the availability of organ biobanking is a bit like trying to envision attempting to distribute human blood with a 24-hour shelf life limitation.

Biobanking of organ and tissue replacements has not been widely discussed perhaps in part because the technology for doing this without damage to the graft is not in hand. Although freezing can achieve limited success for some organs,[7,9,12-15] freezing of the heart, liver or kidney has not been accomplished with subsequent life support function following cooling to temperatures low enough for long-term preservation, despite work on this problem dating

* Reprinted by permission from ***Organogenesis*** [2009 Jul–Sep; 5(3): 167–175]. Copyright © 2009 Landes Bioscience.

back to the 1950s.[3,6] Kidneys and hearts have been the most widely studied organs, but neither has been reproducibly recovered after freezing to temperatures lower than about −20°C,[16-20] evidently due at least in part to mechanical damage from ice itself,[21-24] although in the case of kidneys at least, sporadic survival has sometimes been claimed after freezing to about −40 to −80°C.[25-28]

Some time ago, one of us (GMF), after witnessing transplanted dog kidneys turning deep blue and passing urine that resembled whole blood after freezing to only −30°C with 3 M glycerol (unpublished observations using the same methodology[29] used for rabbit kidney freezing), proposed a way of cooling organs to cryogenic temperatures without incurring the consequences of ice formation.[30-33] This is possible because high concentrations of cryoprotective agents reduce the likelihood and the speed of ice crystal formation, and sufficiently high concentrations can prevent ice formation completely, even at the low cooling and warming rates that are applicable to organ-sized objects.[1,34-36] Cooling an ice-free biological system to a low enough temperature eventually results in a transition from a mobile fluid state to a molecularly arrested glassy state (this transition being referred to as vitrification, or the glass transition). A glass is essentially a liquid that cannot flow over most time scales of interest to the observer,[36] and a vitrified biological system can theoretically be stored for virtually any desired length of time due to the extreme slowing of all diffusion-driven change below the glass transition temperature[37] (T_G). "Vitrification solutions"[38] are solutions of cryoprotective agents that are sufficiently concentrated to enable vitrification or virtual vitrification of a living system at the cooling rates employed for that purpose.

Major advances in vitrification technology have recently been reported,[6,39] and it is now possible to vitrify entire organs,[6] but to do so with full recovery of viability after transplantation is still difficult due in large part to devitrification. Devitrification is ice formation during rewarming, and it arises because ice nuclei, which form initially only at temperatures too low for appreciable crystal growth,[36,40] encounter temperatures during warming that maximize ice growth.[40,41] To date, small ovaries,[42-45] blood vessels,[11] heart valves,[46] corneas[47] and similar structures[48] that can all be cooled and rewarmed rather rapidly so as to avoid devitrification, are the only macroscopic structures that have been reported to recover at least in part after vitrification.

Research on vitrification of organs that require immediate vascular anastomosis upon transplantation has been carried out primarily on the rabbit kidney.[5,6,39,49-51] The rabbit kidney provides a useful illustration of the general problems of preserving both natural and laboratory-generated organ replacements. In this article, we describe the special problems of vitrifying the kidney and progress made toward their solution, including the first case of life support after vitrification and rewarming.

§2. Results

Survival of the first large solid organ after vitrification and transplantation: a case history.

In late 2002 and early 2003, several rabbit kidneys were perfused with the M22 vitrification solution,[6] vitrified and transplanted[52] back to their original donors (autografts) with immediate contralateral nephrectomy either to evaluate survival or to evaluate short-term blood reflow only for the first several minutes in vivo. No rabbit survived when perfused with M22 at 40–60 mmHg, but one of two survived after perfusion with M22 at 80 mmHg for 25 min, and the second rabbit in this small group lived for 9 days after transplantation, which was longer than any other non-surviving rabbit studied. Although anecdotal, the sole survivor proves that organ cryopreservation by vitrification can result in life-supporting function after transplantation, and a detailed examination of this case reveals many interesting aspects of the problem of successfully preserving an organ by vitrification.

The events during perfusion of the surviving kidney are shown in **Figure 1A** and, with the exception of the elevated perfusion pressure during M22 perfusion, are typical of protocols we have described for several years.[5,6,49] The venous concentration just before cooling the kidney to below T_G was 96.4% of the arterial concentration, and the absolute arteriovenous concentration difference was 330–340 mM. Under the conditions of this perfusion, this venous concentration predicts[6] an inner medullary tissue concentration that is 92.1% of the arterial concentration, which is sufficient to permit vitrification on cooling although insufficient to preclude devitrification.[6]

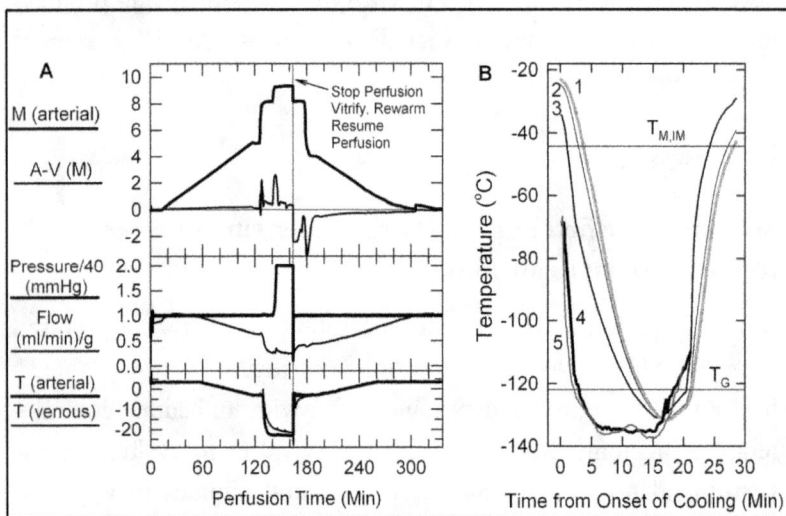

Figure 1.

A. Perfusion protocol for renal survival after vitrification and rewarming. M, molarity; A-V (M), arteriovenous difference in molarity; T, temperature in degrees Celsius. The protocol, as usual,[6, 39] employs an initial 5M plateau, a second plateau at 8.4M to allow cooling to -22°C without freezing, and a final plateau during M22 perfusion. In the experiment shown, the perfusion was interrupted at the point shown to enable the kidney to be vitrified, rewarmed, and reperfused with 8.4M cryoprotectant at -3°C.

B. Thermal history of the transplanted kidney based on invasive temperature measurements in a model rabbit kidney cooled and rewarmed by a procedure identical to that used for the vitrified-transplanted rabbit kidney. Line 1: inner medullary temperature, as documented by a thermocouple located 1.2 cm below the renal surface; line 2: outer medullary temperature, measured 7 mm below the renal surface; line 3: cortical temperature 2 mm below the renal surface; line 4: environmental temperature of the test kidney; line 5: environmental temperature of the kidney that was transplanted after previous vitrification. $T_{M,IM}$, estimated melting point of inner medullary tissue (upper horizontal line); T_G, estimated glass transition temperature of inner medullary tissue (lower horizontal line).

Figure 1B shows the thermal history of the kidney during cooling and warming and indicates that all parts of the transplanted kidney were below T_G for about 8 min, the thermal nadir being about −130°C for the cortex, outer medulla and inner medulla (approximately 7–8°C below the estimated T_G of the inner medulla[6]). The warming rate of the inner medulla from T_G to −60°C was about 15°C/min, and declined to 6°C/min from −60°C to the predicted inner medullary melting point ($T_{M,IM}$) of ~−44.2°C. During the removal of M22, the kidney perfused normally, and during transplantation, the urine was not bloody and the kidney appearance was reasonable and seemed to be recovering at closure.

The animal became anemic on the first postoperative day and again on day 10 (**Fig. 2A**). This symptom was not previously seen after cooling to −45°C.[6] Fortunately, the anemia spontaneously resolved after being successfully treated, suggesting recovery of adequate renal production of erythropoietin. Acute hyperkalemia developed on days 2 and 3 but was successfully controlled. Thereafter, K^+ levels slowly rose until reaching a stable value by about day 32. Serum creatinine peaked at 14.6 mg/dl (**Fig. 2B**) on day 4 and then fell to a nadir of 3.3 mg/dl on day 24 independent of diuresis and hydration. It then slowly rose again until reaching an apparently stable value of 6.0–6.4 by day 38.

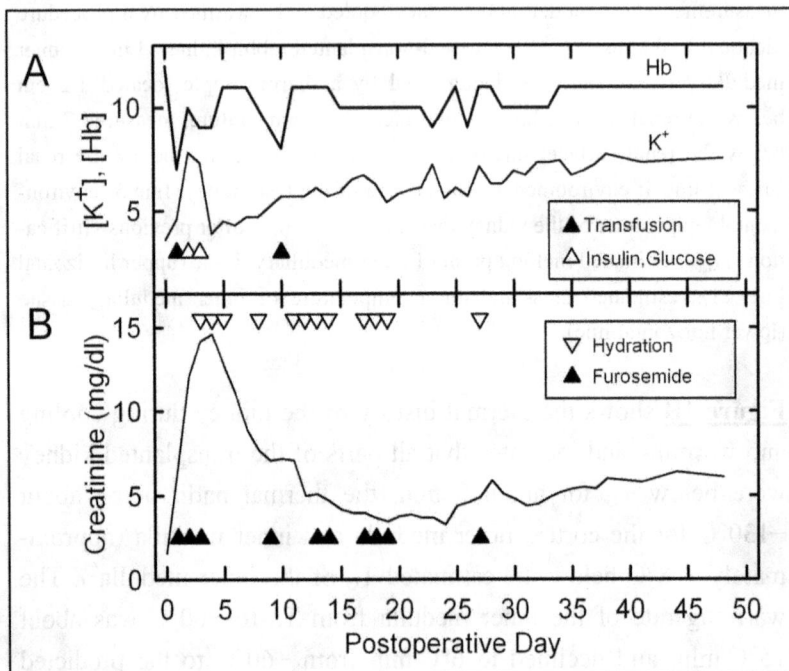

Figure 2.

A. Changes in blood levels of hemoglobin and potassium after transplantation of a previously vitrified rabbit kidney and interventions to correct both (triangles). Hyperkalemia was corrected by intravenous glucose (20 ml of 5% dextrose in 0.45% NaCl) and insulin (0.4 ml of 1 U/ml, IV). Anemia was corrected with 20 ml of whole rabbit blood (~6-8 ml/kg) on each occurrence. Blood levels were measured before corrective interventions given on the same day. [Hb], hemoglobin concentration in g/dl; [K^+], potassium concentration in meq/l.

B. Postoperative creatinine levels and diuretic support history. Lower triangles indicate furosemide administration (generally 5-10 mg, IV or IM); upper triangles indicate hydration (generally 100-200 ml, consisting of equal volumes of 0.9% NaCl and 0.45% NaCl plus 5% glucose, subcutaneously). Blood levels were measured before corrective interventions given on the same day.

Clinically, the animal regained normal drinking behavior, a normal fecal output score, and a normal urine volume output score by about 1–2 weeks postoperatively, but food consumption and to a

lesser extent water consumption and urine output declined on bal-
ance after day 24. The rabbit lost about 18% of its body weight by
the fifth postoperative day and thereafter maintained this weight
while also maintaining normal posture and behavior other than
some sluggishness.

After ensuring that the animal appeared capable of living indefinitely using the vitrified kidney as the sole renal support, it was
euthanized for histological follow-up on day 48. Ice formation during warming was not expected in the cortex but was expected to be
equivalent to 1–2% of the total inner medullary mass,[6] so the fate of
the renal medulla was of special interest. To our surprise, examination of an entire renal cross section showed that medullary damage
was essentially confined to one side of the kidney, the medullary
portion of the peripelvic columns on the opposite side displaying
remarkably good survival (**Fig. 3**). This raises fascinating but still
unresolved mechanistic questions about the origin of the observed
damage, but indicates that under the experimental conditions
achieved, the delivery of M22 to the medulla was sufficient to allow survival of considerable medullary mass, inspiring hope that
relatively small improvements in medullary cryoprotectant delivery
might enable full survival of the renal medulla.

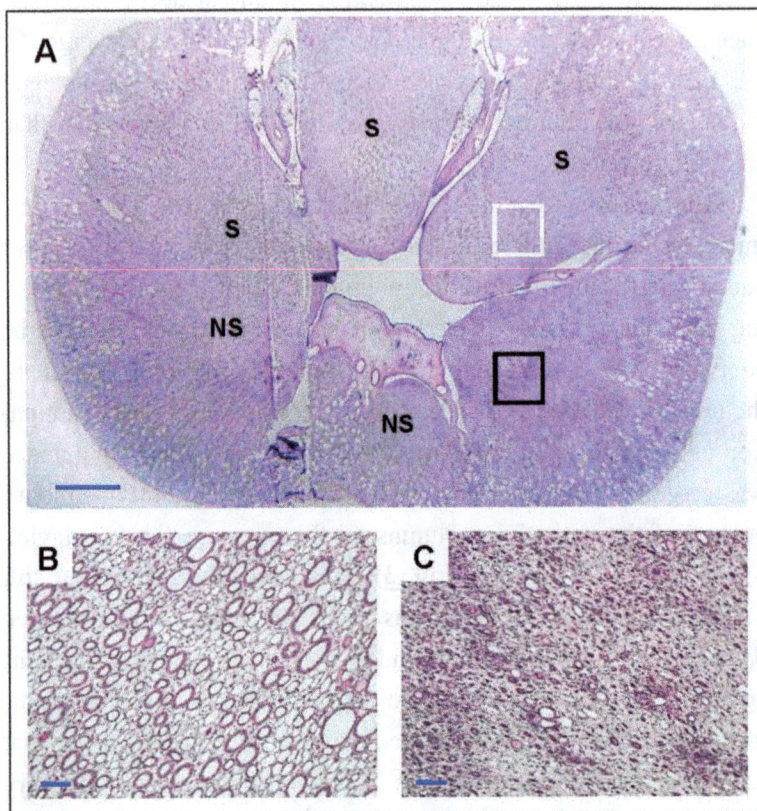

Figure 3.
A. Cross-section of the vitrified/rewarmed kidney (PAS staining) showing surviving (S) and non-surviving (NS) medullary areas; white box designates the region depicted in Figure 3B, and black box identifies the location of Figure 3C. Non-surviving areas are confined to one side of the kidney. Scale bars: in A, 3 mm; in B and C, 100 microns.

Despite lack of expected freezing in the renal cortex, considerable cortical injury was observed as well. This damage ranged from reasonably mild loss of superficial cortical tubules (**Fig. 4**, top) to predominant loss of tubules in the cortex corticis with persistence of glomeruli (**Fig. 4**, middle) to loss or atrophy of both superficial tubules and associated glomeruli (**Fig. 4**, bottom). We speculate that this injury is the result of previously undiscovered stress-strain phenomena in the outer cortex caused by the establishment of large thermal gradients in relatively stiff and brittle tissue near the glass transition temperature. Lowering cooling or warming rates to avoid this form of injury is feasible in principle but will require still better distribution of M22 into the renal medulla because medullary devit-rification will otherwise be exacerbated by lower cooling and warming rates, as verified by direct observation (unpublished re-sults).

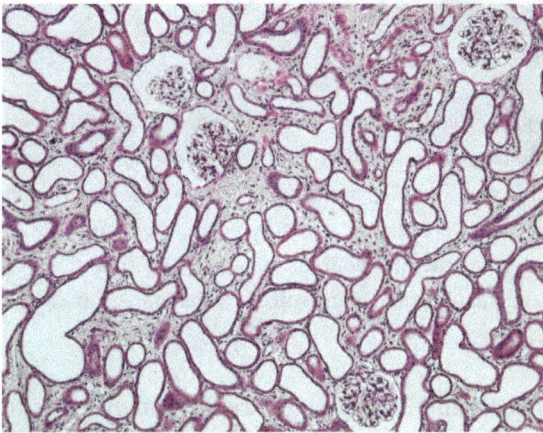

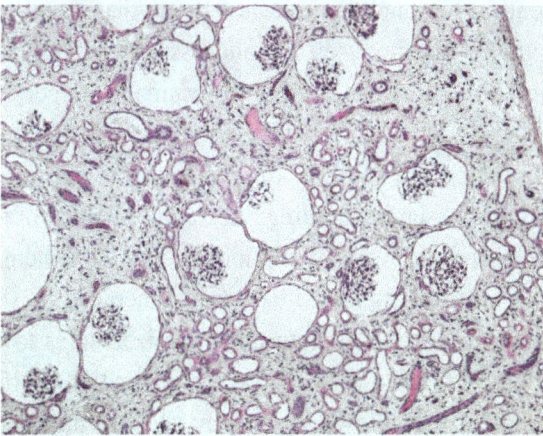

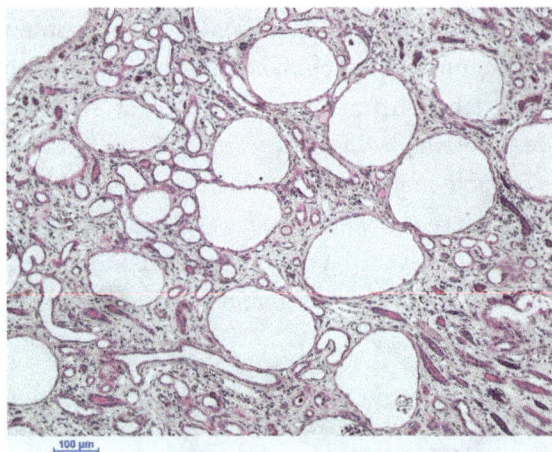

Figure 4.

The spectrum of renal cortical responses to vitrification and rewarming. Top: area showing predominant survival of both tubules and glomeruli. Scale bars all represent 100 microns. Middle: transitional zone between predominantly surviving superficial renal cortex and non-surviving cortex, showing loss of tubules but survival of glomeruli. Bottom: non-surviving superficial cortex, showing loss of both tubules and glomeruli, with ballooning of Bowman's capsule. PAS stain.

The problem of renal medullary water replacement.

These results identify the renal medulla as a tissue that seems to be poised at the dividing line between the success and the failure of vitrification. The survival of the medulla presumably depends on the relationship between medullary cryoprotectant delivery and medullary ice formation, and deeper insight into this relationship will be fundamental for understanding the requirements for successful vitrification and recovery of the kidney and, by extension, for the recovery of vitrified organized tissues in general.

The anatomy of the renal vasculature is organized so as to constrain medullary blood flow to a small fraction of total renal blood flow,[53] an arrangement that allows the kidney to concentrate urine but makes the task of delivering cryoprotectant to the medulla a difficult one. Anatomically, the medullary circulation is provided by the vasa recta, which originate either directly from widely-spaced points along the arcuate arteries or indirectly from efferent arterioles of juxtamedullary glomeruli, which comprise about 9% of the total number of glomeruli;[54] in either case, the originating blood vessels subdivide into many parallel vascular channels, each of which carries a small fraction of the flow that enters the originating vessel.

These anatomical limitations are a given, but medullary delivery of cryoprotectant can be influenced by factors such as perfusate viscosity, cryoprotectant delivery protocol, and the permeability and diffusivity of the cryoprotectants in the vitrification solution. In addition, the vascular system is not the only route of delivery for cryoprotectants. The medulla consists also of tubules and collecting ducts that can convey permeable cryoprotectants along their lengths, diffusing as they go. At the temperatures of our experiments[6] (−22°C to 3°C, and particularly ~−22°C or −3°C for delivery of the highest concentrations of cryoprotectants), and in the presence of more than 8 molar cryoprotectant (<~48% v/v water), no appreciable renal metabolism can be expected, and therefore tubular delivery of cryoprotectants to the medulla is presumably entirely passive and driven only by filtration at the glomerulus followed by local diffusion (no secretion, no active reabsorption, just diffusion) until delivery into the pelvis. Given that medullary blood flow amounts to only about 10% of total renal blood flow under ordinary conditions,[53] a filtration fraction in the vicinity of just 10%, which we have observed for rabbit kidneys,[49] would be sufficient to deliver enough ultrafiltrate to the medulla to match the total volume flowing through the medullary blood vessels.

The best vitrification solution known for the kidney to date is M22,[6] whose critical cooling rate (the cooling rate above which ice formation is not observed) is 0.1°C/min, and whose critical warming rate (the warming rate above which ice formation is not observed) is 0.4°C/min.[1,34] As determined from the cooling and warming curves of **Figure 1B**, the rabbit kidney can, conservatively, be cooled and warmed by conduction at about 8°C/min or more, which implies that 100% equilibration of the medulla with M22 is not required. Key questions are, what is the level of equilibration that is required, what is required to achieve it, and how can we know when we have achieved it? These questions are taken up in the next section.

Measuring and achieving adequate equilibration.

Comparisons between cryoprotectant concentrations in the urinary space and in the venous effluent revealed that the "urine" (perfusate ultrafiltrate) tends to lag far behind the venous effluent in concentration (**Fig. 5**). This is logical since the urine flow rate is a fraction of the arterial flow rate, and since the venous effluent disproportionately samples the overperfused renal cortex, which accounts for ~90% of total renal perfusate flow, and therefore under-represents poorly-equilibrated areas. In addition, the urine makes three passes through the renal medulla (descending and ascending limbs of the loop of Henle followed by passage through the collecting ducts) and therefore is in intimate osmotic/diffusive communication with the renal medulla before it is collected. For these reasons, the urine is expected to reflect medullary tissue concentrations of cryoprotectant better than is the venous effluent concentration, and experimental results described below bear out this expectation.

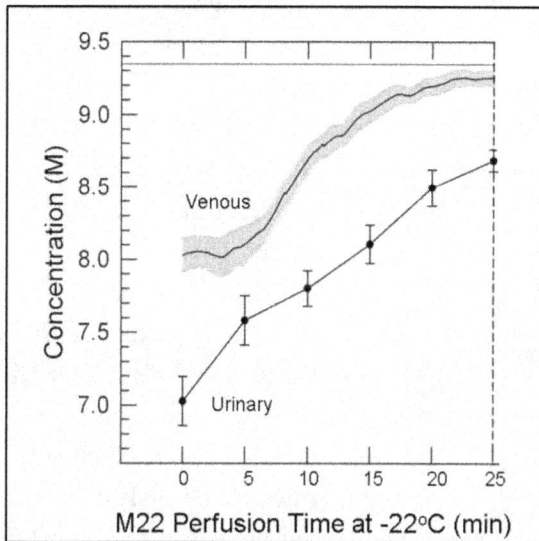

Figure 5.
Difference between the venous concentration and the urinary space concentration during M22 perfusion at -22°C and 40 mmHg (n = 4 perfusions). Urine concentrations (discreet data points ± 1 SEM) determined manually; venous concentrations (line with gray "halo" consisting of ± 1 SEM) determined by computer. The time base gives time from the nominal onset of M22 perfusion, which includes a lag time as M22 makes its way through the perfusion circuit. The horizontal line near the top of the graph shows the concentration of M22, which is not fully reached even by the venous effluent by the end of M22 delivery.

Figure 6 provides a basis for illustrating many features of medullary cryoprotectant introduction. The figure shows the effects of perfusion temperature and the polymer content of M22 on arterial flow and urine concentration equilibration with the arterial perfusate. As in **Figure 1A**, perfusion with the VMP transitional solution[39] (8.4 M total concentration, the second concentration plateau of **Fig. 1A**) begins at −3°C and in the standard protocol[6] continues during continuous perfusion-cooling to −22°C to allow M22 perfusion to begin at −22°C, but we see in **Figure 6** that when this is done (M22 -22), the urine concentration lags so far behind the arterial concentration that at the end of M22 perfusion, urine concentration is just reaching the concentration of VMP, or only about 90% of the full concentration of M22. Perfusion of VMP and M22 only at higher temperatures (−3°C) reduces viscosity and greatly improves both arterial flow and equilibration, as expected, allowing about 95% equilibration to be attained (M22 -3). Removing all polymers from M22 at −3°C (M22NP) further reduces viscosity, improves flow, and improves equilibration, as expected. However, an anomaly is introduced when M22NP is supplemented with just one of the polymers of M22, namely, the commercial Supercool X-1000™ ice blocker[55] (X-1000). Perfusing this solution (M22NP + 2X, containing 2% w/v X-1000, or twice the usual concentration of X-1000 in M22,[6]) slows the arterial perfusion rate yet ultimately allows a degree of equilibration similar to that achieved after M22NP perfusion. Therefore, urinary space equilibration is not proportional to the arterial flow rate.

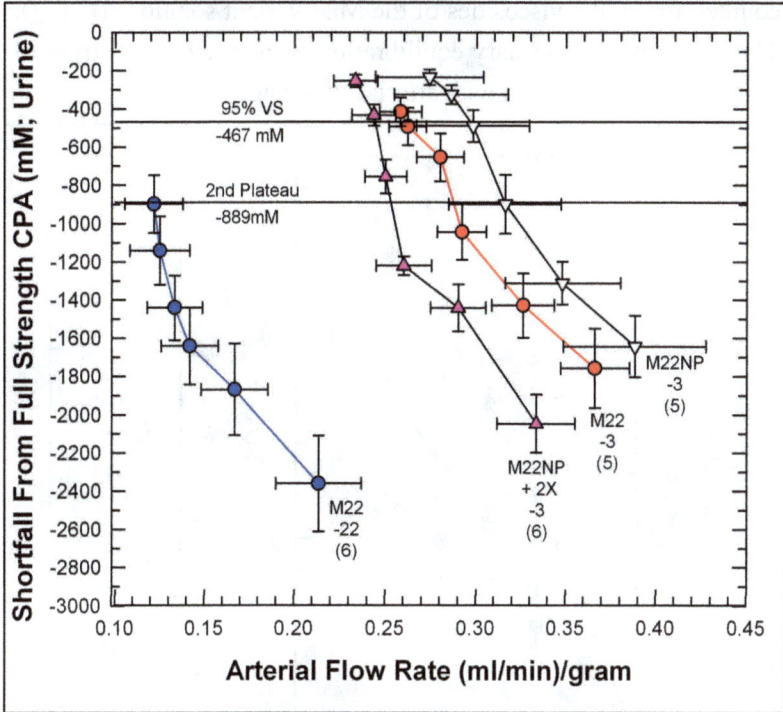

Figure 6.

Equilibration shortfalls (urine concentration minus nominal arterial concentration) in rabbit kidneys perfused with M22 at -22°C (M22 -22) or at -3°C (M22 -3) plotted as a function of arterial flow rates (which decline as higher concentrations are reached and viscosity increases). M22NP -3 refers to M22 minus all polymers, perfused at -3°C; M22NP + 2X -3 refers to M22NP containing 2% X1000 ice blocker, perfused at -3°C. Values in parentheses indicate the number of perfusions of each type. Each data point represents urine equilibration measured at 5-min intervals, beginning at VS perfusion time zero to the right and ending at VS perfusion time = 25 min to the left. The horizontal lines are "landmark" concentrations and refer to the concentrations of VMP (2nd Plateau, which falls at a shortfall of -889 mM) and 95% of full-strength vitrification solution (VS) (which, because of the negligible molarity of the polymers of M22, is essentially the same for M22, M22NP, and M22NP+2X). Error bars designate ± 1 SEM. For discussion, see text.

Equilibration was, however, mirrored by differences in the urine flow rates in these groups, and the latter were in turn closely accounted for by the viscosities of the M22 variant solutions (**Fig. 7**). Thus, it seems that urinary equilibration is more closely correlated with urine flow rate than with arterial flow rate.

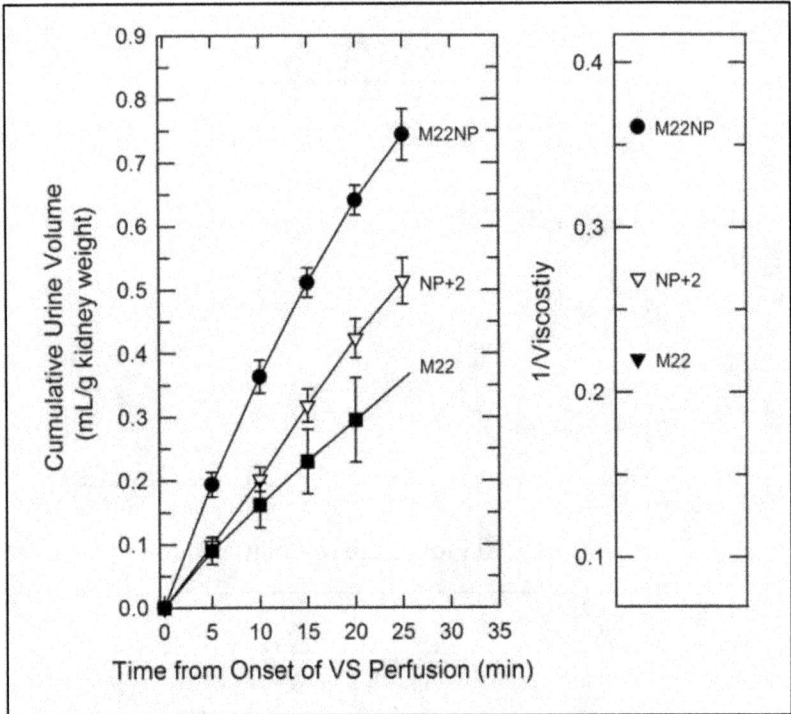

Figure 7.
Left: urine accumulation during perfusion with M22, M22NP + 2% X1000 (NP + 2), and M22NP at -3°C; right: reciprocal viscosities of these three vitrification solutions (cP[-1]). The total accumulated urine volumes are inversely proportional to the total viscosity of each VS (M22 = 4.54 cP; M22NP + 2X = 3.71 cP; M22NP = 2.77 cP). The urine volume for M22 at 25 min was not consistently recorded and so is indicated by extrapolation. Data points represent means ± 1 SEM.

Figure 8 answers the question of "how much equilibration is enough" and brings out a number of other important points. The left panels describe the devitrification temperatures, percent ice formed at the point of devitrification, and percent of ice melted at the tissue melting point, for urine samples collected at the end of the perfusion, and the right panels report the same information for inner medullary tissue samples (all data obtained by differential scanning calorimetry).

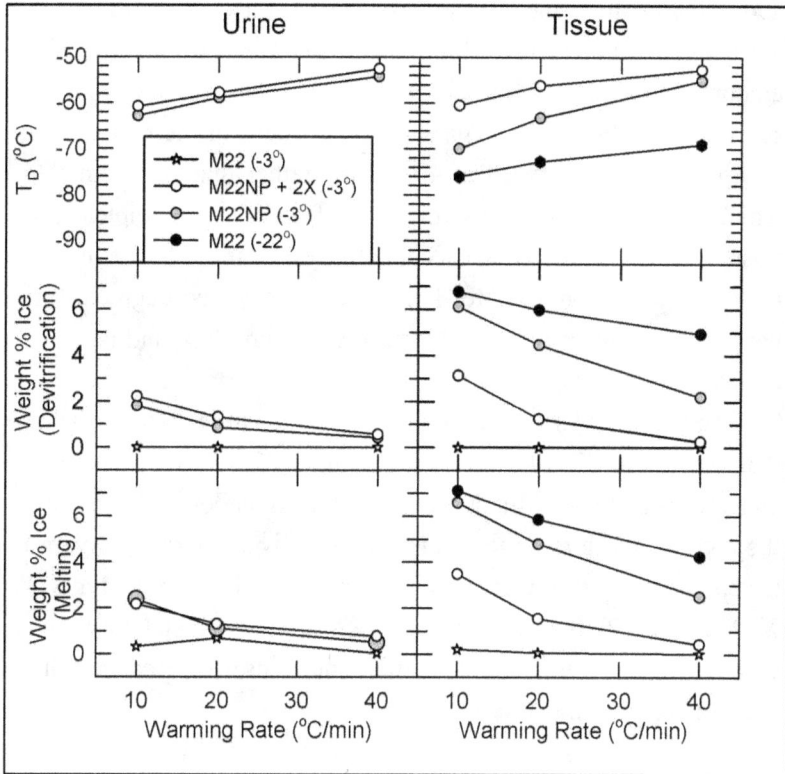

Figure 8.

Temperature, extent, and warming rate dependence of ice formation in urine (left panels) and tissue samples (right panels) obtained from kidneys subjected to the four protocols of Figure 6 and relationship between the amount of ice formed during devitrification and the amount of ice that thawed upon complete rewarming. Urine was not collected from the M22 kidneys perfused at -22°C. Upper panels: devitrification temperatures (T_D); middle panels: the percentage of sample mass that crystallizes during devitrification; lower panels: the percentage of sample mass that melts upon continued warming. Each point represents the mean of generally 5-6 independent measurements; devitrification temperatures are averaged only for those samples that devitrified. No devitrification event was observed for any specimen in the M22 -3°C group.

Error bars omitted for clarity. Groups are represented as indicated in the inset. For discussion, see text.

The first thing to note is that perfusion of M22 at −22°C causes about 7% of inner medullary mass to crystallize as ice during rewarming, and this result is little affected by the warming rate. This amount of ice is substantially greater than was predicted for our surviving vitrified kidney, presumably because in the experiments of **Figure 8** we perfused at 40 mmHg rather than at 80 mmHg, which is known to make a significant difference.[6] In complete contrast, perfusion of M22 at −3°C (stars) results in no tissue ice formation at any warming rate. Therefore, the required degree of urinary space equilibration lies between 90% and 95%, and is probably close to the latter limit.

Second, perfusing M22NP at −3°C results in less ice formation than perfusing M22 at −22°C even though M22NP is a more dilute and intrinsically less stable solution; this is undoubtedly because the higher equilibration level of M22NP delivers more net cryoprotectant despite its lower total concentration. Finally, adding 2% X1000 to M22NP greatly suppresses tissue ice crystal formation, which demonstrates the ability of X1000 to usefully penetrate into and protect inner medullary tissue.

Comparing tissue results to urine results shows that tissue generally devitrifies at a lower temperature than does urine from the same kidney, and that the amount of ice formed in tissue is accordingly higher than it is in urine from the same kidney, indicating that tissue concentrations lag behind urine concentrations. Interestingly, for both urine and tissue, in most cases the percentage of sample mass that melts upon thawing is the same as the percentage that freezes during devitrification, meaning that vitrification is generally complete on cooling with the regimen used for tissue analysis.

Visual assessment of ice formation.

Although tissue biopsies allow quantitative results to be obtained as presented in **Figure 8**, we have been interested in developing methods for visualizing ice formation across entire renal cross-sections in order to be able to judge the two and three-dimensional extent of ice formation. Although these methods are still in development, we present an example of the type of information that can be obtained in **Figure 9**. In this example, the warming rate was about 1°C/min, and therefore more ice is expected than with the more rapid warming used in **Figure 8**. Nevertheless, the maximum extent of ice formation, judged by whitening of the tissue during rewarming, did not include the cortex in this example, and the ice that formed appeared to be uniformly distributed.

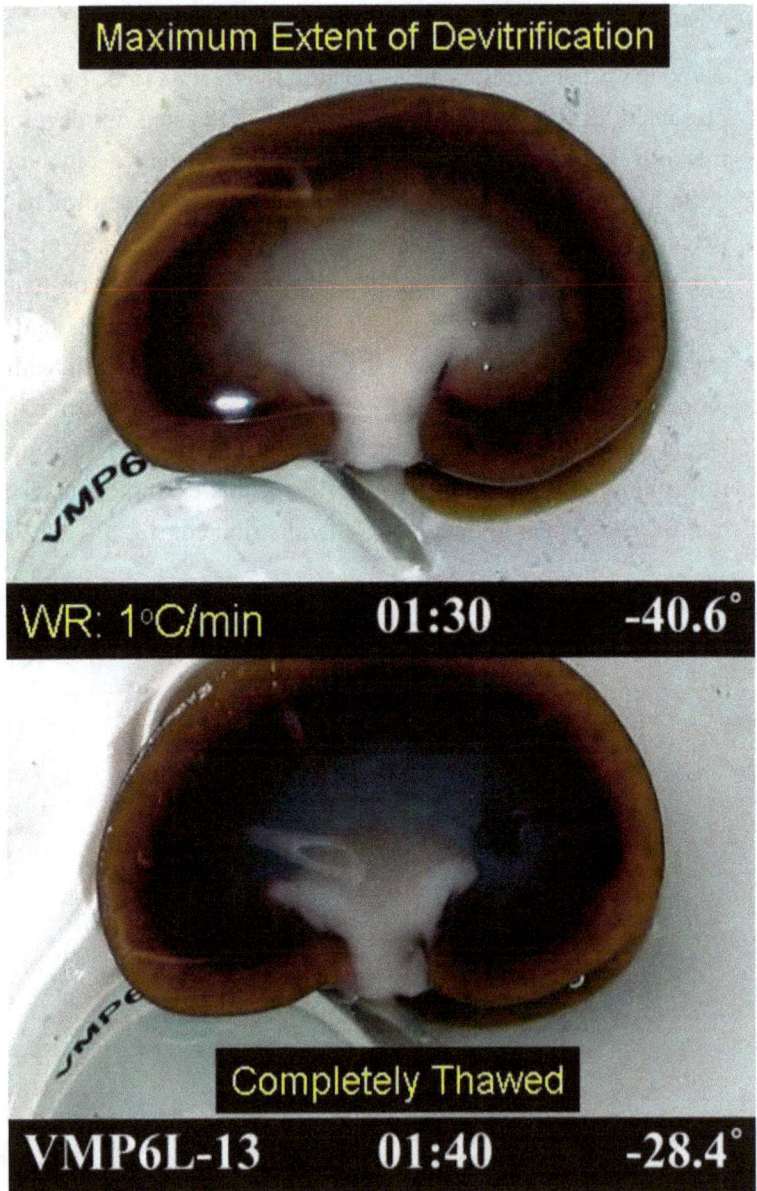

Figure 9.
Visual appearance of ice in an exemplary rabbit kidney cross-section during rewarming. The kidney was perfused with M22 at -22°C, cut in half, immersed in M22, vitrified in a CryoStar freezer at -135°C, and eventually rewarmed at about 1°C/min while being photographed from time to time. Re-

warming was accomplished by transfer of the kidneys to an insulated box through which liquid nitrogen vapor was circulated slowly so as to allow steady warming of the contained atmosphere from just below Tg to well above the renal melting points. Times (1:30 and 1:40) represent times in hours and minutes since the onset of slow warming, and temperatures refer to ambient atmospheric temperatures near the kidney but not within the kidney itself. The upper panel shows the kidney at the point of maximum ice cross-sectional area, and the lower panel shows the kidney after complete ice melting. Both panels show the site of an inner medullary biopsy taken for differential scanning calorimetry.

Using this method and differential scanning calorimetry will eventually allow us to determine the extent to which medullary ice formation can be tolerated by the kidney. We have been able to show that medullary damage can be assessed in the acute postoperative period by removing kidneys 30 min after transplantation, flushing them to remove blood, and examining the extent of medullary blood trapping by inspection of renal cross-sections (unpublished observations). Although preliminary, such observations have identified conditions that allow blood trapping to be avoided, and as our methods improve, we should be able to use such methods to determine how much medullary ice formation, if any, is acceptable, and to select perfusion methods for evaluation by permanent transplantation.

§3. Materials and Methods

Procedure for obtaining survival after rabbit kidney vitrification.

A 12.7 gram rabbit kidney was perfused with M22, a 9.3 M vitrification solution with very low critical cooling and warming rates,[1,6,34] in an LM5 carrier solution under computer control[56] using a variation of our standard protocol[6] on December 10th, 2002 (**Fig. 1A**). Perfusion began with Renasol-14 containing 2% w/v B. Braun hydroxyethyl starch (HES) and no cryoprotectant and continued, after a pause at 5 M cryoprotectant to allow the arteriovenous (AV) concentration gradient to level, to VMP[39] in LM5 containing no HES. To distribute M22 more thoroughly than usual while minimizing damage from perfusion pressures over 40 mmHg, perfusion pressure was raised to 80 mmHg only during the 25-min period of exposure to M22 itself. The kidney was removed from the perfusion apparatus at the end of M22 perfusion and cooled in rapidly-moving nitrogen vapor[6] (**Fig. 1B**) The intra-renal thermal history was determined by inserting a three-point needle thermocouple (beads at 2, 7 and 12 mm depths; PhysiTemp, Huron, PA) into an identically-treated but non-transplanted kidney.

Rewarming was accomplished by slowly raising the environmental temperature to about −115°C in order to bring the cortical temperature to just above T_G, at which point the kidney was returned to the perfusion machine and further warmed by pouring M22 at −22°C over the renal surface for 8 min. Rewarming was completed by perfusing the kidney with VMP at −3°C, after which cryoprotectant washout was completed as usual[6] (**Fig. 1A**), transitioning from VMP in LM5 to Renasol-14 + 2% HES, and the kidney was transplanted according to our published method[52] with immediate contralateral nephrectomy.

Perfusion of kidneys with M22 and alternative vitrification solutions at 40 mmHg.

All perfusions were carried out under computer control in the general manner represented in **Figure 1A**. However, because B. Braun discontinued the manufacturing of HES and because the use of all alternative forms of HES was associated with higher post-transplantation peak creatinine levels (unpublished results), we replaced HES with 2% w/v decaglycerol (dG) in the carrier solution at the beginning of the perfusion. We also used TransSend-4 at the beginning and end of each experiment, but retained LM5 as the carrier for VMP, M22, and their variants. Remaining protocol details other than the perfusion pressure were as reported in **Figure 1A** and elsewhere.[6]

End point measurements.

Determination of tissue freezing points (devitrification temperatures), percent ice formation and percent ice melted were all carried out by differential scanning calorimetry. The cooling and warming protocol for inner medullary samples was to cool to $-120°C$ at $10°C/min$ and to rewarm at 10, 20 or $40°C/min$ when the endpoint was devitrification. Heats of devitrification and of melting were obtained by integrating peak areas and were converted from units of joules/gram into percent ice formation by dividing by 3.34.[6] The temperatures of devitrification were taken to be the temperatures at the tops of the observed peaks. Cryoprotectant concentrations were determined from refractive index readings on the basis of appropriate calibration curves. Baseline data were freshly derived for each experiment during priming of the perfusion system. All refractive indices were recorded continuously at ~0°C using ice-immersed in-line process refractometers (AFAB Enterprises, Eutis, FL, Model PR-111) at the beginning (priming) and experimental phases of each perfusion except that urine refractive index was determined using a bench-top Bellingham Stanley RFM 330 refractometer at room temperature and converted to concentration using a separate room temperature calibration obtained using the same refractometer. Viscosities were measured using a Gilmont falling-ball viscometer (Cole Parmer) at room temperature.

§4. Conclusions

Clearly, the problem of eliminating or sufficiently limiting ice formation throughout the kidney without inducing unacceptable toxicity is a complex and many-faceted one. So far, the most promising single approach seems to be the one described in **Figure 1**, which resulted in survival after transplantation. However, the many lessons that have been learned since that experiment will undoubtedly result in methods for protecting the kidney that are more effective than those used in **Figure 1**, and that will allow better and more consistent survival to be obtained after vitrification and rewarming. Certainly, the availability of new methodologies to evaluate renal tissue resistance to ice formation will be helpful, and the use of microwave rewarming to reduce the likelihood of damage from devitrification could also be highly beneficial for our efforts to solve the very complex problem of fully successful renal vitrification.

Because of its unique vascularization, the kidney may be the most challenging organ of them all to vitrify and rewarm successfully. If so, continued progress with the kidney should be encouraging for the future vitrification and recovery of other complex living systems, including laboratory-produced organ and tissue replacements, whose accessibility to cryoprotectant may be significantly greater than that of the renal medulla.

Acknowledgements

Transplantation of the surviving kidney was carried out by Dr. Jun Wu, who is now in private dental practice. The authors wish to thank Ms. Perlie Tam for expert surgical assistance. Supported by 21st Century Medicine, Inc., All procedures involving animal use were done according to USDA standards and with IACUC approval.

References

1. Fahy GM, Wowk B, Wu J. Cryopreservation of complex systems: the missing link in the regenerative medicine supply chain. *Rejuvenation Res.* 2006;9:279–291.

2. Starzl TE. A look ahead at transplantation. *J Surg Res.* 1970;10:291–297.

3. Smith AU. Problems in the resuscitation of mammals from body temperatures below 0°C. *Proc R Soc Lond B Biol Sci.* 1957;147:533–544.

4. Karow AM., Jr . The organ bank concept. In: Karow AM Jr, Abouna GJM, Humphries AL Jr, editors. *Organ Preservation for Transplantation.* Boston: Little, Brown and Company; 1974. pp. 3–8.

5. Khirabadi B, Fahy GM. Permanent life support by kidneys perfused with a vitrifiable (7.5 molar) cryoprotectant solution. *Transplantation.* 2000;70:51–57.

6. Fahy GM, Wowk B, Wu J, Phan J, Rasch C, Chang A, et al. Cryopreservation of organs by vitrification: perspectives and recent advances. *Cryobiology.* 2004;48:157–178.

7. Wang X, Chen H, Yin H, Kim S, Lin Tan S, Gosden R. Fertility after intact ovary transplantation. *Nature.* 2002;415:385.

8. Karlsson JO, Toner M. Cryopreservation. In: Lanza RP, Langer R, Vacanti J, editors. *Principles of Tissue Engineering.* Second Edition. San Diego: Academic Press; 2000. pp. 293–307.

9. Arav A, Revel A, Nathan Y, Bor A, Gacitua H, Yavin S, et al. Oocyte recovery, embryo development and ovarian function after cryopreservation and transplantation of whole sheep ovary. *Hum Reprod.* 2005;20:3554–3559.

10. Kaiser J. New prospects for putting organs on ice. *Science.* 2002;295:1015.

11. Song YC, Khirabadi BS, Lightfoot F, Brockbank KG, Taylor MJ. Vitreous cryopreservation maintains the function of vascular grafts. *Nat Biotechnol.* 2000;18:296–299.

12. Bedaiwy MA, Jeremias E, Gurunluoglu R, Hussein MR, Siemianow M, Biscotti C, et al. Restoration of ovarian function after autotransplantation of intact frozen-thawed sheep ovaries with microvascular anastomosis. *Fertil Steril.* 2003;79:594–602.

13. Martinez-Madrid B, Dolmans M-M, van Langendonckt A, Defrere S, Donnez J. Freeze-thawing intact human ovary with its vascular pedicle with a passive cooling device. *Fertil Steril.* 2004;82:1390–1394.

14. Hamilton R, Holst HI, Lehr HB. Successful preservation of canine small intestine by freezing. *J Surg Res.* 1973;14:313–318.

15. Fahy GM. Analysis of "solution effects" injury: rabbit renal cortex frozen in the presence of dimethyl sulfoxide. *Cryobiology.* 1980;17:371–388.

16. Elami A, Gavish Z, Korach A, Houminer E, Schneider A, Schwalb H, et al. Successful restoration of function of frozen and thawed isolated rat hearts. *J Thorac Cardiovasc Surg.* 2008;135:666–672.

17. Toledo-Pereyra LH. Organ freezing. *J Surg Res.* 1982;32:75–84.

18. Pegg DE, Green CJ, Walter CA. Attempted canine renal cryopreservation using dimethyl sulphoxide, helium perfusion and microwave thawing. *Cryobiology.* 1978;15:618–626.

19. Smith AU. The effects of glycerol and of freezing on mammalian organs. In: Smith AU, editor. *Biological Effects of Freezing and Supercooling.* London: Edward Arnold, Ltd; 1961. pp. 247–269.

20. Kubota S, Lillehei RC. Some of the problems associated with kidneys frozen to −50°C or below. *Low Temp Med.* 1976;2:95–105.

21. Pegg DE, Diaper MP. The mechanism of cryoinjury in glycerol-treated rabbit kidneys. In: Pegg DE, Jacobsen IA, Halasz NA, editors. *Organ Preservation, Basic and Applied Aspects.* Lancaster: MTP Press, Ltd; 1982. pp. 389–393.

22. Karow AM, Jr, Shlafer M. Ultrastructure-function correlative studies for cardiac cryopreservation. IV. Prethaw ultrastructure of myocardium cooled slowly (<=2°C/min) or rapidly (>=70°C/sec) with or without di-methyl sulfoxide (DMSO) *Cryobiology.* 1975;12:130–143.

23. Hunt CJ. Studies on cellular structure and ice location in frozen organs and tissues: the use of freeze-substitution and related techniques. *Cryobiology.* 1984;21:385–402.

24. Pollack GA, Pegg DE, Hardie IR. An isolated perfused rat mesentery model for direct observation of the vasculature during cryopreservation. *Cryobiology.* 1986;23:500–511.

25. Halasz NA, Rosenfield HA, Orloff MJ, Seifert LN. Whole organ preservation II. Freezing studies. *Surgery.* 1967;61:417–421.

26. Halasz NA, Miller S. Rewarming methods for whole organ freezing. In: Norman JC, editor. *Organ Perfusion and Preservation.* New York: Appleton-Century-Crofts; 1968. pp. 731–737.

27. Guttman FM, Lizin J, Robitaille P, Blanchard H, Turgeon-Knaack C. Survival of canine kidneys after treatment with dimethylsulfoxide, freezing at −80°C, and thawing by microwave illumination. *Cryobiology.* 1977;14:559–567.

28. Lehr H. Progress in long-term organ freezing. *Transplant Proc.* 1971;3:1565.

29. Fahy GM. Activation of alpha adrenergic vasoconstrictor response in kidneys stored at −30°C for up to 8 days. *Cryo Letters.* 1980;1:312–317.

30. Fahy GM. Prospects for vitrification of whole organs. *Cryobiology.* 1981;18:617.

31. Fahy GM, Hirsh A. Prospects for organ preservation by vitrification. In: Pegg DE, Jacobsen IA, Halasz NA, editors. *Organ Preservation, Basic and Applied Aspects.* Lancaster: MTP Press; 1982. pp. 399–404.

32. Fahy GM, MacFarlane DR, Angell CA, Meryman HT. Vitrification as an approach to cryopreservation. *Cryobiology.* 1984;21:407–426.

33. Fahy GM, MacFarlane DR, Angell CA. Recent progress toward vitrification of kidneys. *Cryobiology.* 1982;19:668–669.

34. Wowk B, Fahy GM. Toward large organ vitrification: extremely low critical cooling and warming rates of M22 vitrification solution. *Cryobiology*. 2005;51:362.

35. Wowk B, Fahy GM. Ice nucleation and growth in concentrated vitrification solutions. *Cryobiology*. 2007; 55:330.

36. Wowk B. Thermodynamic aspects of vitrification. *Cryobiology*. 2010; 60:11-22.

37. Fahy GM, Rall WF. Vitrification: An overview. In: Liebermann J, Tucker MJ, editors. *Vitrification in Assisted Reproduction: A User's Manual and Troubleshooting Guide*. London: Informa Healthcare; 2007. pp. 1-20.

38. Rall WF, Fahy GM. Ice-free cryopreservation of mouse embryos at −196°C by vitrification. *Nature*. 1985;313:573–575.

39. Fahy GM, Wowk B, Wu J, Paynter S. Improved vitrification solutions based on predictability of vitrification solution toxicity. *Cryobiology*. 2004;48:22–35.

40. Fahy GM. The role of nucleation in cryopreservation. In: Lee REJ, Warren GJ, Gusta LV, editors. *Biological ice nucleation and its applications*. St. Paul: APS Press; 1995. pp. 315–336.

41. Fahy GM. Vitrification. In: McGrath JJ, Diller KR, editors. *Low Temperature Biotechnology: Emerging Applications and Engineering Contributions*. New York: American Society of Mechanical Engineers; 1988. pp. 113–146.

42. Courbiere B, Massardier J, Salle B, Mazoyer C, Guerin J-F, Lornage J. Follicular viability and histological assessment after cryopreservation of whole sheep ovaries with vascular pedicle by vitrification. *Fertil Steril*. 2005;84:1065–1071.

43. Sugimoto M, Maeda S, Manabe N, Miyamoto H. Development of infantile rat ovaries autotransplanted after cryopreservation by vitrification. *Theriogenology*. 2000;53:1093–1103.

44. Salehnia M. Autograft of vitrified mouse ovaries using ethylene glycol as cryoprotectant. *Exp Anim.* 2002;5:509–512.

45. Migishima F, Suzuki-Migishima R, Song S-Y, Kuramochi T, Azuma S, Nishijima M, et al. Successful cryopreservation of mouse ovaries by vitrification. *Biol Reprod.* 2003;68:881–887.

46. Brockbank KG, Song YC. Morphological analyses of ice-free and frozen cryopreserved heart valve explants. *J Heart Valve Dis.* 2004;13:297–301.

47. Armitage WJ, Hall SC, Routledge C. Recovery of endothelial function after vitrification of cornea at −110°C. *Invest Ophthalmol Vis Sci.* 2002;43:2160–2164.

48. Taylor MJ, Song YC, Brockbank KG. Vitrification in tissue preservation: new developments. In: Fuller BJ, Lane N, Benson EE, editors. *Life in the frozen state.* Boca Raton: CRC Press; 2004. pp. 603–641.

49. Fahy GM, Ali SE. Cryopreservation of the mammalian kidney II. Demonstration of immediate ex vivo function after introduction and removal of 7.5 M cryoprotectant. *Cryobiology.* 1997;35:114–131.

50. Fahy GM, da Mouta C, Tsonev L, Khirabadi BS, Mehl P, Meryman HT. Cellular injury associated with organ cryopreservation: chemical toxicity and cooling injury. In: Lemasters JJ, Oliver C, editors. *Cell Biology of Trauma.* Boca Raton: CRC Press; 1995. pp. 333-356.

51. Khirabadi BS, Fahy GM, Ewing L, Saur J, Meryman HT. 100% survival of rabbit kidneys chilled to −32°C after perfusion with 8 M cryoprotectant at −22°C. *Cryobiology.* 1994;31:597.

52. Wu J, Ge X, Fahy GM. Ultrarapid nonsuture mated cuff technique for renal transplantation in rabbits. *Microsurgery.* 2003;23:1–5.

53. Ofstad J, Aukland K. Renal circulation. In: Seldin DW, Giebisch G, editors. *The kidney, physiology and pathophysiology.* New York: Raven Press; 1985. pp. 471–496.

54. Kaissling B, Kritz W. Structural analysis of the rabbit kidney. *Adv Anat Embryol Cell Biol.* 1979;56:1–123.

55. Wowk B, Leitl E, Rasch CM, Mesbah-Karimi N, Harris SB, Fahy GM. Vitrification enhancement by synthetic ice blocking agents. *Cryobiology.* 2000;40:228–236.

56. Fahy GM. Organ perfusion equipment for the introduction and removal of cryoprotectants. *Biomed Instrum Technol.* 1994;28:87–100.

Human Cryopreservation Research at Advanced Neural Biosciences

Aschwin de Wolf & Chana de Wolf

Introduction

In 2008 we obtained modest funding to establish a laboratory aimed at researching cryonics. Our first challenge was to establish a research program that (a) would distinguish itself from other research labs engaged in cryobiology research, and (b) would be feasible in terms of limited financial resources and time. We immediately recognized that our greatest contribution would be to investigate cryonics protocols under realistic conditions. In this article we introduce the reader to some of our most important and robust discoveries.

Until cryonics becomes available as an elective medical procedure, all cryonics patients will experience varying degrees of cerebral ischemia. Even in "good" cases where stabilization procedures are initiated promptly after pronouncement of legal death, the agonal period prior to cardiopulmonary arrest can give rise to cerebral perfusion impairment. In the case of cryonics organizations that do not offer standby and stabilization services, we should expect at least 24 hours of cold ischemia for a typical (remote) patient, often preceded by significant periods of warm ischemia due to no, or slow, cooling.

The fact that no cryonics patient can completely escape some degree of cerebral ischemia forces cryonics organizations to deal with a fundamental question: how do our protocols and vitrification solutions perform under such conditions? In particular, in our lab we have been interested in the behavior of vitrification solutions in ischemic brains. It should not be *a priori* assumed that vitrification solutions preferred for non-ischemic tissues are preferred for ischemic tissues as well. A related line of research is whether the composition of carrier solutions can be altered to improve cryoprotectant perfusion in the ischemic brain.

The investigation of cryonics procedures under realistic conditions is by no means exhausted by conducting experiments under ischemic conditions. Another major difference between cryobiology experiments conducted in the laboratory and the practice of cryonics is that the control over perfusion temperatures is limited in cryonics cases. Even the most sophisticated cryonics protocols expose the brain to toxic concentrations of the vitrification agent at high sub-zero temperatures. Thus, our earliest investigations in 2009 were concerned with the effects of exposing red blood cells to high concentrations of VM-1 (the vitrification agent of the Cryonics Institute) in order to address the possibility that exposing a patient to high concentrations of this agent in the absence of rigorous temperature control could produce instantaneous red blood cell lysis (i.e., hemolysis).

The Red Blood Cell as a Model for Cryoprotectant Toxicity

Various approaches are available to investigate cryoprotectant toxicity, ranging from theoretical work in organic chemistry to cryopreservation of whole mammalian organisms. One simple model that allows for "high throughput" investigations of cryoprotectant toxicity uses red blood cells (erythrocytes). Although the toxic effects of various cryoprotective agents may differ among red blood cells, other cells, and organized tissues, positive results in a red blood cell model can be considered the first experimental hurdle that needs to be cleared before the agent is considered for testing in more advanced models. Because red blood cells are widely available for research, this model eliminates the need for animal experiments for initial screening studies. It also allows researchers to investigate human red blood cells. Other advantages include the reduced complexity of the model (packed red blood cells can be obtained as an off-the-shelf product) and lower costs.

Red blood cells can be subjected to a number of different tests after exposing them to a cryoprotective agent. The most basic test is gross observation of the red blood cells in a cryoprotectant solution. When high concentrations of a cryoprotectant are introduced (such as in vitrification), a stepwise approach is necessary to avoid osmotic damage. If a cryoprotectant solution is extremely toxic rapid hemolysis will follow, which can be observed as a noticeable change of the color of the solution, hemolyzed cell debris sinking to the bottom of the test tube, or negligible difference between the pellet (if there is one at all) and the supernatant after centrifugation. It is important to keep in mind that these effects only indicate *gross* membrane damage and that absence of hemolysis is not equivalent to absence of cryoprotectant toxicity.

In our investigations we did not observe instantaneous hemolysis of sheep red blood cells when 70% of VM-1 (in carrier solution) was introduced in a stepwise fashion either at room temperature or close to the freezing point of water. Morphological studies with light microscopy showed slight alterations for VM-1 (dehydration, decreased uniformity) but we have not seen the extreme alterations and destruction that have been observed in solutions that were formulated to produce hemolysis. Eliminating the step-wise approach and exposing the red blood cells to 70% of VM-1 at once, however, *did* produce hemolysis. This effect was more pronounced at *lower* temperatures, presumably because at low temperatures the rate of diffusion of cryoprotectants is further depressed than the rate of diffusion of water, causing more pronounced osmotic damage.

VM-1 consists of 35% dimethyl sulfoxide (DMSO) and 35% ethylene glycol (EG). Cryobiologist Yuri Pichugin identified this binary cryoprotectant as one of the least toxic (non-patented) binary vitrification solutions for the vitrification of rat hippocampal brain slices. DMSO is a stronger glass former than EG, but in the case of

DMSO as a mono-agent, stepwise exposure of red blood cells to a 70% solution produced complete instantaneous hemolysis. This observation corroborates the contribution of *specific* toxicity to hemolysis of red blood cells and the need for toxicity neutralization in vitrification solutions.

Since red blood cell hemolysis assays are not optimal for quantifying minor differences in cryoprotectant toxicity, or for investigating the effects of cryoprotectants on organized nervous tissue, we limit our use of this method to preliminary investigations of new variants of VM-1 and/or alternative carrier solution composition.

Perfusion of the Ischemic Brain

The brain distinguishes itself from most other organs by its high energy utilization. When the brain is deprived of oxygen and other energy substrates, a complex biochemical cascade ensues that ultimately results in decomposition. Since we do not know the degree of degradation that still permits meaningful reconstruction of the original state of the brain, the most conservative approach is to limit ischemia as much as is practically possible.

The human brain is too large to use immersion as a method to replace water with a cryoprotectant. This fact necessitates the use of vascular perfusion to prepare the brain for exposure to cryogenic temperatures. As a consequence, the ability to protect the brain against ice formation is not an independent challenge but depends on the state of the brain at the time of cryoprotective perfusion. It is at this juncture of ischemia and cryoprotective perfusion where we have conducted most of our experiments.

In a non-ischemic brain, sub-optimal equilibration of the vitrification solution may be compensated by dehydration. This phenomenon is of limited relevance to patients with extensive cerebral ischemia because, as ischemia progresses, the blood-brain barrier through which such dehydration is mediated will become progressively disrupted. For example, cryoprotective perfusion of the non-ischemic rat produces *severe* dehydration of the brain. After 24 hours of cold ischemia, this dehydration is sharply reduced, and after 48 hours there is no evidence of cerebral dehydration after cryoprotectant perfusion. This phenomenon allowed us to investigate cryoprotective perfusion under ischemic conditions without modifications to the carrier solution to limit cryoprotectant-induced shrinking of the brain.

Our first approach to study the effect of ischemia on perfusion impairment in the brain was to add India ink to the perfusate. Areas with no, or poor, perfusion are distinguished by residual blood and absence of ink. In those studies we limited ourselves to investigating the perfusability of the brain *without* subsequent freezing to obtain a basic understanding of this phenomenon without additional variables.

Inspection of the brain after ink perfusion showed that 60 minutes of ischemia at room temperature is sufficient to produce noticeable perfusion impairment with the degree and distribution of the impairment worsening progressively as the duration of warm ischemia increases.

Two interventions that are presumed to mitigate perfusion impairment are antithrombotic therapy and induction of hypothermia. Administration of the anti-coagulant heparin prior to ischemia and the thrombolytic streptokinase following ischemia failed to improve perfusion. This outcome corroborates that ischemia-induced "no-reflow" is not confined to blood clotting and suggests a role for the involvement of blood in a non-coagulating fashion. Scientific and clinical reviews of the no-reflow phenomenon have identified several other factors that contribute to perfusion impairment including red cell aggregation, vasogenic and cellular edema, free radical damage, and inflammatory mediators. Some studies, including the cerebral resuscitation studies of Peter Safar and colleagues, have found benefits from a combination of high perfusion pressures and hemodilution. Our studies into such protocols for short periods of ischemia are inconclusive and for longer (>24 hours) periods of cold ischemia we have found that higher perfusion pressures during cryoprotective perfusion *increase* ice formation after cooling to cryogenic temperatures.

One of our most robust findings is that rapid induction of hypothermia after circulatory arrest mitigates the no-reflow phenomenon. Perfusion impairment was greatly reduced when the brain was cooled *in situ* using a miniature portable ice bath. The whole-body cooling rate in these experiments exceeded 1°C per minute. Since such cooling rates are not practically feasible during external cooling in human cryopreservation stabilization without an aggressive combination of different cooling modalities, including cyclic lung lavage, we repeated these experiments at a cooling rate (~ 0.18 °C per minute) that is practical for human cryopreservation and observed the same benefits. These findings strongly corroborate the current practice of rapid induction of hypothermia in cryonics and suggest that even modest decreases of brain temperature can significantly mitigate perfusion impairment, even if the reduction in metabolic demand cannot prevent exhaustion of energy in the brain.

Another consistent finding in our research is that blood substitution prior to circulatory arrest strongly reduces perfusion impairment. In the India ink model we did not observe evidence of perfusion impairment after up to 72 hours of cold ischemia following blood substitution with m-RPS-2 (the carrier solution of VM-1). One limitation of this model is that complete washout of the blood prior to ischemia excludes observation of residual blood after perfusion as an indicator of perfusion impairment. Filling of vessels with India ink correlates strongly with the degree of perfusion impairment but it does not rule out the presence of small pockets of poorly perfused areas in the brain. Because India ink perfusion may not completely predict the degree of cryoprotectant equilibration that is possible after warm and cold ischemia, we further refined our model and introduced observation of the degree of ice formation after cryoprotectant perfusion and cooling as an endpoint.

Cryoprotective Perfusion of the Ischemic Brain

As a general rule, cryonics interventions aimed at preventing and mitigating ischemic injury are not evaluated with cryoprotective perfusion and ice formation as an endpoint. As a consequence, there is a serious lack of knowledge about the efficacy of cryonics stabilization protocols on reducing ice formation. One of the most valuable research models in our lab has been to conduct cryoprotective perfusion under various conditions of (cold) ischemia. Space limitations prevent us from disclosing all our findings, but our most important discoveries are discussed below. Most of our investigations into cryopreservation of the ischemic brain have been conducted with VM-1, the vitrification solution of the Cryonics Institute.

The most fundamental and robust finding in these experiments is that the duration of warm and cold ischemia is positively associated with perfusion impairment and ice formation after cooling to liquid nitrogen temperatures. Our studies corroborate the pioneering feline work that cryonics researcher Michael Darwin did in this area with electron micrographs in the 1980s. In the rat brain we have identified a consistent hierarchy of vulnerability to cold ischemia-induced perfusion impairment as revealed by inspection of the perfused brain and signs of ice formation after cryogenic cooling. The following four major areas are ranked by increasing vulnerability:

Cerebral cortex; cerebral subcortex; cerebellar cortex; cerebellar subcortex.

We do not have a full understanding of the reason behind this ranking but these findings may be somewhat comforting in light of our current understanding that the most identity-critical information is stored in the cortex of the brain and that the cerebellum may be the least important area in this regard. Notwithstanding this, our research has been aimed at overcoming perfusion impairment and ice formation in patients with extensive ischemic exposure.

We have studied a number of different interventions to improve outcome in cold ischemic brains and the majority of our experiments involved *alteration of the cryoprotectant carrier solution*. We started by adding various non-permeating salts and sugars and high molecular weight polymers to the carrier solution in order to mitigate edema under the expectation that this would improve outcome. This approach did not produce the desired outcome and meaningful reduction of interstitial edema was not observed either.

We did observe improved outcome in terms of reduction of perfusion impairment in the presence of suitable concentrations of the high molecular weight polymers PVP K360, dextran 500, and dextran sulfate 500. We initially attributed these encouraging outcomes to the ability of these polymers to "seal" leaky membranes, although this interpretation seemed to be at odds with the lack of edema reduction observed.

A real breakthrough occurred when we designed a number of solutions that were made *equiviscous* with a dextran sulfate 500 based carrier solution – our most successful carrier solution to date. All these solutions produced comparable results in terms of overcoming perfusion impairment, indicating that the advantageous properties of these higher molecular weight solutions was not specific to their chemical composition but may be mediated through higher viscosity. This interpretation was further corroborated by our observation that we could also produce improved outcome when we conducted cryoprotective perfusion at lower subzero temperatures, which also increases viscosity of the solutions. Protocols that gradually decreased viscosity during cryoprotective perfusion with the aim of taking advantage of the vessel-clearing properties of higher viscosity solutions at the start of perfusion and improved equilibration of the vitrification solution towards the end of perfusion failed to improve upon protocols in which the viscosity was kept constant (for a given pressure) across all steps.

Contrary to what one would expect from the vast literature on the no-reflow phenomenon, conducting cryoprotectant perfusion at high pressures (> 100 mmHg) in brains with 24 and 48 hours of cold ischemia worsened the outcome. We speculate that these high pressures "push" more perfusate with low glass-forming properties into the interstitial space, limiting the equilibration of the higher concentrations of the vitrification solution during later stages of perfusion. As a matter of fact, many of our best results were obtained when we lowered the perfusion pressure below our standard arterial line pressure of 100 mmHg. We also observed improved perfusion and reduced ice formation when we eliminated one or two steps in our three-step perfusion protocol. This finding may offer some important clues to the mechanisms that contribute to improved cryoprotectant perfusion in the ischemic brain. Since starting with such high initial concentrations of the cryoprotectant at the start of cryoprotective perfusion clearly contradicts basic cryobiology practice to minimize osmotic injury and consequent cell rupture, we have not explored this approach in much detail.

So far, we have employed three distinct cooling methods. In our earliest cooling experiments we used liquid nitrogen plunging to cool samples to liquid nitrogen temperatures. To avoid fracturing, we later modified a small lab dewar to allow a more gradual descent of the temperature to -130°C (slightly below the glass transition temperature of VM-1). Currently we employ an ultra-low temperature electrical freezer that can cool samples to -130 degrees Celsius, which also permits us to store our samples for longer periods our time. Our findings concerning ice formation after cryoprotective perfusion of the ischemic brain have been identical for all three cooling methods. The distribution of ice formation generally follows the areas of perfusion impairment observed prior to cooling, which validated the investigations we conducted with India

ink. We have not found any benefits for the addition of pharmacological agents to the carrier solution. Our best understanding about
cold ischemia-induced cryoprotective perfusion impairment is that
two major contributing factors are red cell aggregation (i.e., hyper-
viscosity) and edema.

Organ Preservation Solutions

Remote blood substitution in cryonics has a number of important
(theoretical) arguments in favor of the practice. Replacing the blood
with an organ preservation solution extends the period that organs
can be received from static storage in clinical organ preservation.
The procedure also permits a faster cooling rate in the field than is
possible with external cooling alone. The mannitol-based perfusate
MHP-2 that is currently used by the Alcor Life Extension Foundation has been developed in a series of experiments where dogs were
recovered after 5 hours of asanguineous ultraprofound hypothermia.

   Cryobiology researcher Yuri Pichugin has questioned the value
of remote blood substitution in cryonics because none of the organ
preservation solutions that he tested (including MHP-2 and UW
Solution) could maintain viability of hippocampal brain slices for
periods that are typical of transport times in cryonics practice. Our
own research, however, has been informed by the possibility that
remote blood substitution may fall short in terms of preserving
viability but could still confer benefits in terms of improving cryoprotective perfusion.

   We have compared controls (i.e., no blood substitution) against
the following washout solutions: m-RPS-2, RPS-2 and MHP-2; and
observed that blood substitution does confer significant benefits in
terms of improving cryoprotective perfusion and reducing ice formation. In particular, MHP-2 outperformed the other solutions and
has allowed us to conduct cryoprotective perfusion after 48 hours
of cold bloodless ischemia with no ice formation in the brain after

cooling below the glass transition temperature. Even at 72 hours, ice formation is relatively minor compared to 72 hours of cold ischemia in which the blood is left in the brain, which produces *severe* perfusion impairment and ice formation. These experiments vindicate the practice of remote blood substitution in cryonics, but also emphasize that the *composition* of the organ preservation solution matters a great deal.

None of the organ preservation solutions we have tested (including more advanced recent formulations from colleagues) mitigate the severe vasogenic edema that is observed during cryopreservation after prolonged periods of cold ischemia. We have designed a number of experiments to improve upon the formulation of MHP-2 but none of these variants has been successful so far in decreasing edema and frequently produced worse results than MHP-2 in reducing ice formation after bloodless cold ischemia.

Cryopreservation *after* Chemical Fixation

The idea to chemically fix the brain prior to cryopreservation has remained a topic of interest among cryonics advocates. As a matter of fact, this procedure was discussed in Eric Drexler's classic treatment of molecular nanotechnology, *Engines of Creation*. One argument that could be offered in favor of this procedure is that it halts the development of ischemia in patients with long expected delays between pronouncement of legal death and cryopreservation. For a long time this idea has been met with skepticism because of (unpublished) experimental observations that such protocols risk producing intracellular freezing during cooling. Because the current generation of cryoprotectants is designed to eliminate ice formation altogether we revisited this topic and designed experiments to study the effects of cryopreservation after chemical fixation.

When there is *no ischemic delay* prior to chemical fixation, chemical fixation still permits cryoprotective perfusion, and no ice formation in the brain was observed after cooling to liquid nitrogen temperatures after up to two weeks of hypothermic storage of the fixed brain *in vivo*. These experiments have been unique in that no whole body edema was observed during cryoprotective perfusion. We did, however, observe *severe* dehydration of the brain following cryoprotective perfusion of the fixed brain, a phenomenon we were not able to eliminate when we added an agent to open the blood brain barrier to our carrier solution.

A practical limitation of cryopreservation after fixation is that delays between pronouncement of legal death and fixation could compromise the efficacy of this procedure and produce the kind of freezing damage that has traditionally been associated with this procedure. When we delayed chemical fixation by an hour, washout of the blood and fixation were incomplete and extensive ice formation followed cryoprotective perfusion. This phenomenon may be overcome by alteration of the fixative carrier solution and different perfusion protocols, but it is doubtful that such sophisticated protocols can be realized in most of the cases where the combination of chemical fixation and cryoprotection may be attractive.

Electron Microscopy of the Ischemic Brain

In collaboration with Dr. Michael Perry of the Alcor Life Extension Foundation we have prepared brain tissue samples for electron microscopy for time points up to 81 hours of normothermic ischemia. Since the rat brain cools at a much faster rate than the human brain after circulatory arrest, we decided that using an incubator to keep the *in vivo* brain at body temperature would be a better and more conservative approximation of what would be expected to occur in human brains. The electron micrographs have given us insight into the ultrastructural properties of the brain after various periods of warm ischemia. Dr. Perry is using these images to develop an algorithm that models the state of ischemic tissue after various periods of warm ischemia.

Dr. Perry has also supported investigations to examine the degree of fixation and long-term effects of delayed fixation of the brain. Preliminary results of these experiments indicate that even short delays between circulatory arrest and chemical fixation of the brain produce incomplete fixation and risk of progressive decomposition of poorly fixed areas over time. Whether such findings discredit chemical fixation as a low cost alternative to cryonics cannot be conclusively resolved by experimental research due to our incomplete understanding of the neuroanatomical basis of identity and the capabilities of future cell repair technologies. One might also argue that a straight freeze is preferable to chemical fixation but that chemical fixation is still preferable to complete decomposition.

Implications for Cryonics Protocols

To date, our investigations into cryopreservation of the ischemic brain strongly support the practice of standby and stabilization in cryonics. In particular, rapid induction of hypothermia after pronouncement of death and remote blood substitution with an organ preservation solution can limit the degree of perfusion impairment and ice formation after cryoprotective perfusion and cooling. We have identified some emerging principles for alteration of carrier solutions and cryoprotective perfusion protocols that can overcome no-reflow in the brain after cold ischemia and reduce ice formation. In patients with varying levels of ischemia, such protocols are still confined to the experimental stage until ultrastructural and viability assays have validated the use of these solutions and protocols.

Our research suggests that chemical fixation of the brain prior to cryoprotective perfusion could be beneficial in case of prolonged (transport) delays, but adverse effects of ischemia limit the use of such protocols to a very narrow set of circumstances in which there is negligible delay between circulatory arrest and chemical fixation.

Future Developments

Future developments in our lab concern further refinements of perfusion and cooling protocols. In our more recent experiments we have been conducting cryoprotective perfusion using an open ramp system that gradually introduces the vitrification agent to the brain (as opposed to distinct steps of increasing concentration) combined with cooling to just below the glass transition temperature of the vitrification solution. We will keep upgrading our cryoprotective setup to make it conform to conventional perfusion equipment; ultimately, we hope to introduce computer controlled features. We also aim to alter our circuit to conduct cryoprotective perfusion at controlled high subzero temperatures.

A major portion of our time and resources in the coming years will be devoted to developing a set of viability assays that can be used to screen the toxicity of improved vitrification solutions. Such assays will not be confined to *in vitro* brain slice work but will include whole brain *in situ* electrophysiology as well.

We have also received financial support to develop a whole body resuscitation model, which will allow us to validate organ preservation solutions and vitrification solutions at hypothermic and high subzero temperatures.

Our effort to simulate realistic cryonics conditions in our lab remains a work in progress. So far we have mostly limited ourselves to cryoprotective perfusion after *either* warm or cold ischemia, with a strong emphasis on cold ischemia. Recent observations in our lab indicate that there is a distinct pathophysiology associated with warm ischemia (and *hyper*thermia) that limits simplistic extrapolations between cold and warm ischemia using the Arrhenius Equation.

In a more realistic cryonics model variable periods of warm ischemia precede cold ischemia. In particular, we aim to investigate the efficacy of blood substitution when blood substitution is delayed; a scenario that is common in cryonics practice and that basically constitutes the rule for organizations that do not offer standby and stabilization services.

Advanced Neural Biosciences, Inc., was incorporated in 2008 and conducts neural cryobiology research. We have received funding and equipment from the Immortalist Society, the Life Extension Foundation, the Cryonics Institute, and the Alcor Life Extension Foundation. We are extremely grateful to Alan Mole, Mark Plus, York Porter, Ben Best, Jordan Sparks, Luke Parrish, James Clement, and Dr. Peter Gouras for additional financial, logistical, and general support.

Cryonics: Introduction and Technical Challenges

Ben Best

INTRODUCTION

In this paper I will attempt to provide both an introduction to cryonics technology and to address some of the technical challenges faced by those attempting to practice cryonics. I begin by defining a couple of terms:

Cryonics is the practice of preserving legally dead humans and animals at cryogenic temperatures in the hope that future science can restore them to healthy living as well as *rejuvenate* them so that they can live hundreds or thousands of years in a healthy, youthful condition.

A key term in this definition is the word *rejuvenate*. Most people are elderly when they die, and would not be so enthusiastic about living thousands of years with the infirmities of old age. Many people don't conceive of rejuvenation when they hear of cryonics, and for that reason fail to see the potential of cryonics. It must be continually stressed that the prospect of rejuvenation is part of what motivates and inspires the desire for cryonics.

Cryogenics is a branch of physics (or engineering) that studies the production of very low temperatures (below -100°C, -148°F) and the behavior of materials at those temperatures.(Sometimes only below 100 Kelvin).

The upper limit for what can be defined as a **cryogenic** temperature can vary from -100°C to 100 Kelvin (-173°C), depending on the author.

Cryonics cannot be called a science insofar as it is based on expectations about the capabilities of future technology. However plausible those explanations may be, they can never be proven. The expectations of future technology seem very plausible to cryonicists. Those expectations include the idea the future medicine will be able to cure all diseases, rejuvenate people, and repair damage incurred during the process of cryopreservation. Cryonics is an ambulance to the future, which is why the term "patients" rather than "corpses" is used for legally dead humans who have been cryopreserved.

In the first stages the circulation and respiration of the cryonics subject is mechanically restored, the subject is administered protective medicines and is rapidly cooled to a temperature between 10°C and 0°C. The subject's blood is washed out and a significant amount of body water is replaced with a cryoprotectant mixture to prevent ice formation. The subject is cooled to a temperature below −120°C and held in cryostasis. When and if future medicine has the capability, the subject will be re-warmed, cryoprotectant will be removed, tissues will be repaired, diseases will be cured, and the subject will be rejuvenated (if required).

EFFECTS OF COOLING

Preservation of food in refrigerators and freezers is based on the principle of lowering temperature to reduce the rate of biochemical degradation. Cooling to reduce metabolic rate (and ultimately to bring chemical processes to a virtual halt) is at the heart of cryonics practice. Initial cooling after pronouncement of death involves placing the cryonics subject in a bath of ice water. Cardiopulmonary support with mechanical active compression/decompression also speeds cooling because of heat transfer from flowing blood.

Cryonics subjects are cooled with convection, a combination of conduction and fluid motion. In convection, a solid object (such as a cryonics subject) is cooled by a fluid (liquid or gas) that is rapidly circulated, such that the fluid can carry heat away from the conduction layer around the solid object. In cooling a cryonics subject from human body temperature (37°C) to 10°C, cooling by rapid circulation of ice-water is far more effective because of the convection effect than cooling by ice-packs or by standing water.

Reduction in temperature can considerably extend the time without blood flow before irreversible damage occurs. Many people, especially children, have been reported to survive 20 minutes to an hour or more of cardiac arrest with complete neurological recovery after hypothermic accidents, such as drowning in cold water[1,2]. Metabolic rate can be dramatically reduced by cooling.

The relationship between reaction rate (**k**) of chemical reactions (including metabolism and the processes of ischemic injury) and temperature (**T**) can be described by the Arrhenius equation[3]:

$$k = Ae^{-E_a/RT}$$

where **T** is in Kelvins, E_a is the **activation energy**, **R** is the **universal gas constant** (8.314 Joules/mole-Kelvin) and **A** is the **frequency factor** (related to frequency of molecular collisions and the probability that collisions are favorably oriented for reaction). Comparing two reaction rates gives the equation:

$$k_1/k_2 = e^{(E_a/R)*(1/T2 - 1/T1)}$$

The reaction rates of enzymes at various temperatures give a close approximation to the relationship between temperature and metabolic rate. Lactate dehydrogenase from rabbit muscle – which has an activation energy (E_a) of 13,100 calories/mole[4] – can be taken as a representative enzyme. Using one thermochemical calorie equal to 4.184 Joules gives 54,810 Joules/mole.

Comparing the reaction rate (k_1) for lactate dehydrogenase at 40°C (313 Kelvins) (T_1) to the reaction rate (k_2) at 30°C (303 Kelvins) (T_2) gives:

$$k_1/k_2 = e^{((54,810\ J/mol)/(8.314\ J/mol\text{-}K))*(1/303\ K - 1/313\ K)} = \mathbf{2.004}$$

This exponential drop in reaction rates with declining temperature means that reaction rates would become infinitesimally small at cryogenic temperatures (temperatures below −100°C) if chemical reactions were possible at those temperatures.

If lactate dehydrogenase reaction rate was representative of metabolism in general, the metabolism at 37°C would be 18 times faster than at 0°C. Experimentally it has been observed that the rate of oxidative phosphorylation at 4°C is about one-twentieth the rate at 37°C[5], a figure roughly in agreement with the value just calculated.

A chemical reaction goes 18 times faster at 37°C than at 0°C, 400,000 times faster than at -80°C (dry ice sublimation temperature), and 10^{28} times faster than at -196°C (liquid nitrogen boiling temperature).

CRYOPROTECTION AND VITRIFICATION

Water molecules are attracted by hydrogen bonds which actually cause the molecules to separate more when a crystal is formed. Ice occupies a 9% greater volume than liquid water, which is why tissues are crushed when ice is formed. Freezing is also damaging because it is associated with high concentrations of toxic electrolytes and – if cooling is fast – osmotic damage.

Quartz (crystalline silicon dioxide) has a crystal structure that can be interrupted by sodium ions, resulting in an amorphous glass. Glass used for windows and beverage containers is made from silicon dioxide that has been made amorphous by the addition of soda (Na_2O) and lime (CaO). Hot glass is a syrup that becomes increasingly viscous upon cooling – a solidified liquid rather than a crystal.

Similar to silicon dioxide, water can form an amorphous solid rather than a crystal when cryoprotective agents are added. The cryoprotective agents EG, PG, glycerol, and DMSO all hydrogen-bond with water – which would interfere with the hydrogen bonding between water molecules that is the basis of ice crystal formation. The addition of cryoprotectants to water to completely eliminate ice formation is called vitrification. Major vitrification achievements include vitrification of a rabbit brain with no ice formation seen anywhere in an electron micrograph[6], hippocampal slices cooled to -130°C and rewarmed with complete viability[7], and rabbit kidney vitrified to -135°C, rewarmed, and able to sustain the rabbit indefinitely as the sole functioning kidney when transplanted into a rabbit[8].

Although cryoprotectants can vitrify, they are toxic chemicals. Cryoprotectant toxicity is the single greatest obstacle preventing suspended animation by cryopreservation. Perfusability is another obstacle. Kidneys are the most difficult organ to vitrify because 98% of blood flow goes to the cortex, whereas only 2% goes to the medulla. But cryoprotectant toxicity is by far the greatest obstacle.

Cryoprotectants become less toxic as temperature declines. Faster cooling rates allow for the use of less concentrated cryoprotectant solutions. Larger organs are harder to cool quickly, which is why it has been possible to vitrify a rabbit kidney, but not a human kidney. Mixtures of cryoprotectant chemicals can be less toxic than pure cryoprotectant chemicals. Ice blockers can be added to cryoprotectant solutions to reduce the amount of cryoprotectant required.

LEGAL DEATH IS NOT ULTIMATE DEATH

As recently as the 1950s it was believed that death is irreversible when the heart stops. Today it is established that CardioPulmonary Resuscitation (CPR) in combination with Automated External Defibrillators (AEDs) can restore many people to life who were clinically dead because of cardiac arrest[9]. But it is still widely believed that after about six minutes of cardiac arrest without circulation irreparable brain damage has already occurred.

In 1976 Peter Safar (the "father of CPR") showed that dogs could be subjected to twelve minutes of cardiac arrest without neurological damage by the use of elevated arterial pressure, norepinephrine, heparin, and hemodilution with dextran 40^{10}. The six minute limit is not mainly a neurological phenomenon, it is a problem of increased vascular resistance that can be overcome (in part) by increasing perfusion pressure[11].

Although legally dead, cryonicists do not believe that cryonics patients are dead in an ultimate sense. Body tissues take quite some time to decompose after the heart stops, which is why cryonicists say "Death is a process, not an event".

CRYONICS PROCEDURES AND ORGANIZATIONS

A cryonics case performed under good conditions will have the following sequence of events:

- Begin cryonics procedures only after death pronounced
- Give medications (especially heparin against clotting)
- Restore circulation/respiration while cooling in ice bath
- CPS CardioPulmonary Support (not CPR, no resuscitation)
- Replace blood with vitrification solution
- Cool quickly (several hours) to just above -123°C
- Cool slowly (several days) to -196°C
- Store in liquid nitrogen (-196°C)
- Future reanimation and rejuvenation of the patients ?

Four cryonics organizations in the world do storage of cryonics patients in liquid nitrogen: Alcor in Arizona, Cryonics Institute in Michigan, KrioRus in Russia, and Trans Time in California. As of October 31, 2010 Alcor and the Cryonics Institute had 101 patients in liquid nitrogen, KrioRus had 12 patients, and Trans Time had 2 patients. 19 of the Cryonics Institute patients are being stored for the American Cryonics Society, a cryonics organization that is only providing services through contractors.

Suspended Animation, Inc. in Florida and EUCrio in Portugal are only devoted to the initial stages of standby, stabilization, and transport of a cryonics patient to other organization for vitrification perfusion and liquid nitrogen storage. "Standby" means waiting by the bedside of a terminal cryonics patient, whereas "Stabilization" means Cooling, CPS and administration of medications after pronouncement of death. Alcor sometimes uses Suspended Animation, Inc. to provide standby/stabilization services, and Cryonics Institute Members can contact with Suspended Animation, Inc. for those services. EUCrio began in 2010, and plans to perform the same services in Europe that are provided by Suspended Animation, Inc. in the United States. Alcor and the Cryonics Institute both perfuse their patients with vitrification solution to prevent ice formation.

Alcor gives its Members the option of cryopreserving the whole body, or of storing only the head at reduced cost. Many cryonicists believe that it is only necessary to store the head, because the memories and personal identity are in the brain, and future technology will be able to recreate whole bodies – bodies better than any human bodies existing today. The Cryonics Institute will only preserve the whole body, no "neuro" option is allowed because of public relations concerns.

When Alcor perfuses the whole body, vitrification solution of the brain hemispheres cannot be monitored independently from the body, and it may take longer for the whole body to be perfused than would be the case if only the head were being pefused. The Cryonics Institute always perfuses the head, but will perfuse the body upon request. If the body is perfused by the Cryonics Institute, glycerol rather than vitrification solution is used because the latter would cause too much edema. Perfusing the body with glycerol causes delay in head cooling, and carries the risk of glycerol seepage into the brain under some circumstances.

Once cardiac arrest has occurred and death has been pronounced, a cryonics subject can be given medications to maintain sedation, reduce cerebral metabolism, prevent/reverse blood clotting, increase blood pressure, stabilize pH against acidosis, and protect against ischemia/reperfusion injury.

Cryonics procedures involve restoring blood circulation and respiration as soon as possible to keep tissues alive. In cryonics, this is called CardioPulmonary Support (CPS) rather than Cardio-Pulmonary Resuscitation (CPR) because resuscitation after death has been pronounced is not desired (a DNR, Do Not Resuscitate, condition). Propofol (2,6-diisopropylphenol) is given partly because its sedative action can prevent resuscitation, with the added benefit that it can be neuroprotective[12]. Propofol has been shown to inhibit the neural cell apoptosis that can occur as a consequence of ischemia/reperfusion injury[13].

Heparin is used to prevent blood clotting. Streptokinase is the usual thrombolytic used to break up blood clots. THAM (Tris-Hydroxymethyl AminoMethane) is a buffer that maintains arterial pH without producing carbon dioxide and also maintains intracellular pH because it readily crosses cell membranes[14].

When the equipment is available, cryonics teams restore circulation and respiration with mechanical devices capable of restoring circulation on the down-stroke (compression) as well as the up-stroke (decompression). Active Compression-DeCompression (ACDC) and interposed abdominal compression can improve CPS perfusion considerably[15]. Epinephrine has commonly been used to supplement CPS by maintaining blood pressure, although vasopressin may also be used[16]. Once the cryonics subject is cooled to below 10°C perfusion with vitrification solution can begin. After vitrification perfusion, the patient is cooled to liquid nitrogen temperature and then stored in liquid nitrogen.

PERFUSION, ISCHEMIA, AND EDEMA

For the sake of clarification, some terms that have already been used should be explicitly defined:

Perfusion: Injection and/or circulation of fluids through blood vessels

Diffusion: Movement of molecules in liquids due to random motion and concentration gradients

Edema: Swelling of body tissues with fluid

Ischemia: Absence of nutrient and oxygen to tissues when there is no blood flow.

The longer the period of ischemia, the more damaged blood vessels become, causing them to become increasingly leaky. Ischemia can result in edema during perfusion due to leaky blood vessels. Warm ischemia applies to cases where there is no cooling after cardiac arrest, whereas cold ischemia describes the condition a cryonics patient experiences when being shipped in ice to a cryonics facility.

Restarting circulation after an extended period of ischemic time can cause more harm than good. Oxygen is beneficial to tissues under normal conditions, but as an increasing amount of time passes without circulation, oxygen increases its capacity to cause free-radical damage to tissues rather than support metabolism[17].

As blood vessels become increasingly damaged by ischemia or reperfusion, the tissues become increasingly edematous upon vitrification perfusion. More fluid enters the body than exits. The tissues can become so swollen with fluid that blood vessels are crushed such that no blood flow occurs. Washing the blood out as part of the stabilization process can improve perfusability and reduce edema if prompt, but can have the opposite effect if delayed, because of reperfusion injury.

Aschwin and Chana de Wolf have done research for the Cryonics Institute through their company Advanced Neural Biosciences. They have discovered ways of modifying vitrification solutions to improve vitrification solution saturation of the brain, despite edema and "no reflow" due to cold ischemia. They can get good brain perfusion after 24 or 48 hours of cold ischemia. They are weekend researchers who could do more for cryonics Technology if there were sufficient funding for them to do cryonics research full-time.

CIRCLE OF WILLIS AND PERFUSION TECHNIQUE

Most blood flow to the brain is through the carotid arteries, but some blood goes through the vertebral arteries. The carotid arteries and the vertebral arteries unite in the Circle of Willis. At least two studies have shown, however, that fewer than 50% of people have a complete Circle of Willis[18,19].

Before Alcor perfuses the head, the head is separated from the body. Perfusion is through the carotid arteries unless perfusate is not seen dripping from the vertebral arteries when perfusion begins, in which case the vertebral arteries would be cannulated and perfused. If dripping from the vertebral arteries is seen, that is taken as evidence that the Circle of Willis is complete – and the vertebral arteries are both clamped for the rest of the perfusion. Perfusing the head in this manner allows for independent Monitoring of vitrification solution saturation for each hemisphere – as determined by the refractive index of the effluent coming from each jugular vein.

Alcor places holes in the scull (burr holes) to allow monitoring of brain shrinkage (or edema). Probes from a crackphone are also placed in the burr holes, which allows for monitoring of cracking during the cooling process that occurs after perfusion. Vitrification solution does not cross the blood brain barrier very well in patients who have not suffered much ischemic damage. In such patients, vitrification is mainly by dehydration (removal of brain water).

Alcor considers the brain to be vitrified after the brain has shrunk to 78% of its original volume.

Out of 15-20 Alcor neuro patients perfused by the above method, all showed dripping from the vertebral arteries, and none were perfused through the vertebral arteries in addition to the carotid arteries. Could it be that the Circle of Willis was complete in every Alcor case? That seems unlikely. A study in which the Circle of Willis was incompete in 60% of subjects found no evidence of insufficient perfusion in functional tests using unilateral perfusion[20]. The explanation for this phenomenon was that there must be extra-cranial collateral circulation.

Prior to 2005, the Cryonics Institute perfused the head and the body with glycerol through the carotid arteries. There was no perfusion through the vertebral arteries. Since 2005 the Cryonics Institute has been ensuring perfusion through both the carotid and vertebral arteries. But the above evidence suggests that perfusing through the carotids would be adequate. Against this idea is the possibility of inadequate perfusion pressure due to backflow through the vertebral arteries.

VITRIFICATION SOLUTIONS

Alcor's vitrification solution is called **M22**, whereas the Cryonics Institute's vitrification solution is called **VM-1**. Both of these vitrification solutions have a solidification temperature of about -123°C upon cooling. Both have isotonic carrier solutions. M22 costs about $170 per liter, whereas VM-1 costs about $1 per liter. VM-1 is a stronger vitrification mixture than M22, but it is also more toxic. The formulas for each of the vitification solutions is given below, along with the explanation of their names:

M22 (optimal perfusion temperature -22°C: Alcor)
22% dimethyl sulfoxide
13% formamide

17% ethylene glycol
3% N-methylformamide
4% 3-methoxy-1,2-propanediol
3% PVP K12
2% Z-1000 ice blocker
1% X-1000 ice blocker

VM-1 (Vitrification Mixture one: Cryonics Institute)
35% dimethyl sulfoxide
35% ethylene glycol

VM-1 is made with industrial grade ethylene glycol and DMSO, but these are very pure from the point of view that the major contaminant is water.

The DMSO used by the Cryonics Institute is 99.7% pure with the only impurities listed being water, "color" and titratable acid (0.001 milliequivalents/gram). The ethylene glycol used by the Cryonics Institute is 99.94% pure with the major impurity again being water: nearly 0.06%. The next largest impurities in the ethylene glycol are acetic acid (<0.001%) and ash (0.0005%). There are part-per-million amounts of chloride, "color", diethylene glycol, and iron in the ethylene glycol.

CRACKING OF VITRIFIED CRYONICS PATIENTS

Cryonics patients are traditionally stored in liquid nitrogen (-196°C), but vitrified tissue solidifies upon cooling at -123°C. For a solid that is cooling from -123°C to -196°C there will be thermal stress because low thermal conductivity will mean that the outside contracts more than the inside. If a sample is small enough, and is cooled slowly enough, cracking due to thermal stress can be avoided. Dr. Brian Wowk cooled a vitrified rabbit kidney from -123°C to -196°C over two days without cracking. But the volume of a rabbit kidney is only 10 milliliters. It could take years to cool a cryonics patient from -123°C to -196°C without cracking.

If thermal conductivity could be increased, cooling a cryonics patient without cracking could be possible. Cryonics patients have traditionally been cooled slowly from -123°C to -196°C over a period of days, in the hope that cracking can be reduced by the slow cooling – even though it is known that cracking cannot be eliminated.

Research reported in 1990 concerning cooling of propylene glycol solutions to cryogenic temperatures demonstrated that cracking is more severe (cracks are finer and more numerous) when it begins to occur at lower temperature than when it begins to occur at higher temperature[21].

As caveats, it should be noted that propylene glycol vitrifies at -109°C, rather than the -123°C solidification temperature seen forthe vitrification solutions used in cryonics. And thermal stress of vitrified biological tissues is only about 75% as great as it is for the vitrification solutions used for vitrification[22]. Neither of these considerations means that cracking can be avoided for cryonics patients.

Based on the 1990 studies, the Cryonics Institute revised its cooling protocol so as to cool rapidly between -118°C to -145°C to force larger cracks to occur at higher temperatures, rather than allow finer cracks at lower temperatures.

An idea for avoiding cracking is to store cryonics patients at -140°C rather than at -196°C (Intermediate Temperature Storage, ITS). A system for implementing ITS was described by Dr. Brian Wowk at a 2007 Suspended Animation, Inc. conference. ITS equipment has been built, but it is not being used. ITS storage would be more expensive than conventional liquid nitrogen storage.

But cracking should not be as difficult for future technology to repair as freezing damage. Moreover, lower cryogenic temperatures show greater protection against radiation damage[23,24]. Eight-cell mouse embryos showed no effect on survival or development during 5-8 months of liquid nitrogen storage despite being subjected to the equivalent of 2,000 years of background gamma-ray radiation[25].

NANOTECHNOLOGY AT CRYOGENIC TEMPERATURE

There has been considerable question about how nanomachines could be powered, especially those operating at cryogenic temperatures. For patients who have been straight-frozen warming above freezing temperatures will immediately give "mush" – just like thawing frozen strawberries (as so many of our ignorant critics like to point out). You will actually start to get "mush" well below freezing temperatures because salt solutions turn liquid well below freezing temperature. Nanobots needing a liquid environment in which to operate create the paradox that as soon as a liquid environment becomes present, broken tissues are subject to hydrolysis and dissolution, if not chemical reactions.

How can tiny robots (nanobots) move at cryogenic temperature? How can there be energy to make and break chemical bonds? How can nanobots be powered at cryogenic temperature? How can billions of nanobots communicate wirelessly?

Ralph Merkle and Robert Freitas have offered some suggested solutions to these problems[26]:

Diamondoid molecular machines (nanobots) could operate at cryogenic temperature. Nanobots could be powered by electrostatic motors. 80% of brain is water or cryoprotectant – nanobots can drill through without damaging structure. Cryogenic repair can take years before other repair at higher temperature (no hurry). Nanobots would communicate with external computers which could do massive computation.

Now it appears that there are "nano-swimmers" – nanowires powered by external magnetic fields swimming like bacteria with flagella[27]. Admittedly, these are operating in solution rather than boring through solids, but at least it shows the potential.

As a last resort, destructive analysis of the brain could be made, digitized, and used to reconstruct a brain from new material. Microtome slicing and recording of brain structure at cryogenic temperature. Duplicates could be made.

REFERENCES

1. Eich C, Brauer A, Kettler D. Recovery of a hypothermic drowned child after resuscitation with cardiopulmonary bypass followed by prolonged extracorporeal membrane oxygenation. Resuscitation. 2005 Oct;67(1):145-8.

2. Bolte RG, Black PG, Bowers RS, Thorne JK, Corneli HM. The use of extracorporeal rewarming in a child submerged for 66 minutes. JAMA. 1988 Jul 15;260(3):377-9.

3. CHEMISTRY:The Central Science (Seventh Edition); pp.511-512; Brown,TL, et. al., Editors; Prentice Hall (1997).

4. Low PS, Bada JL, Somero GN. Temperature adaptation of enzymes: roles of the free energy, the enthalpy, and the entropy of activation. Proc Natl Acad Sci U S A. 1973 Feb;70(2):430-2.

5. Dufour S, Rousse N, Canioni P, Diolez P. Top-down control analysis of temperature effect on oxidative phosphorylation. Biochem J. 1996 Mar 15;314 (Pt 3):743-51.

6. Lemler J, Harris SB, Platt C, Huffman TM. The arrest of biological time as a bridge to engineered negligible senescence. Ann N Y Acad Sci. 2004 Jun;1019:559-63.

7. Pichugin Y, Fahy GM, Morin R. Cryopreservation of rat hippocampal slices by vitrification. Cryobiology. 2006 Apr;52(2):228-40.

8. Fahy GM, Wowk B, Pagotan R, Chang A, Phan J, Thomson B, Phan L. Physical and biological aspects of renal vitrification. Organogenesis. 2009 Jul;5(3):167-75.

9. Cobb LA, Fahrenbruch CE, Walsh TR, Copass MK, Olsufka M, Breskin M, Hallstrom AP. Influence of cardiopulmonary resuscitation prior to defibrillation in patients with out-of-hospital ventricular fibrillation. JAMA. 1999 Apr 7;281(13):1182-8.

10. Safar P, Stezoski W, Nemoto EM. Amelioration of brain damage after 12 minutes' cardiac arrest in dogs. Arch Neurol. 1976 Feb;33(2):91-5.

11. Shaffner DH, Eleff SM, Brambrink AM, Sugimoto H, Izuta M, Koehler RC, Traystman RJ. Effect of arrest time and cerebral perfusion pressure during cardiopulmonary resuscitation on cerebral blood flow, metabolism, adenosine triphosphate recovery, and pH in dogs. Crit Care Med. 1999 Jul;27(7):1335-42.

12. Adembri C, Venturi L, Tani A, Chiarugi A, Gramigni E, Cozzi A, Pancani T, De Gaudio RA, Pellegrini-Giampietro DE. Neuroprotective effects of propofol in models of cerebral ischemia: inhibition of mitochondrial swelling as a possible mechanism. Anesthesiology. 2006 Jan;104(1):80-9.

13. Polster BM, Basanez G, Young M, Suzuki M, Fiskum G. Inhibition of Bax-induced cytochrome c release from neural cell and brain mitochondria by dibucaine and propranolol. J Neurosci. 2003 Apr 1;23(7):2735-43.

14. Gehlbach BK, Schmidt GA. Bench-to-bedside review: treating acid-base abnormalities in the intensive care unit - the role of buffers. Crit Care. 2004 Aug;8(4):259-65.

15. Babbs CF. CPR techniques that combine chest and abdominal compression and decompression: hemodynamic insights from a spreadsheet model. Circulation. 1999 Nov 23;100(21):2146-52.

16. Wenzel V, Lindner KH. Arginine vasopressin during cardiopulmonary resuscitation: laboratory evidence, clinical experience and recommendations, and a view to the future. Crit Care Med. 2002 Apr;30(4 Suppl):S157-61.

17. Zweier JL, Talukder MA. The role of oxidants and free radicals in reperfusion injury. Cardiovasc Res. 2006 May 1;70(2):181-90.

18. Krabbe-Hartkamp MJ, van der Grond J, de Leeuw FE, de Groot JC, Algra A, Hillen B, Breteler MM, Mali WP. Circle of Willis: morphologic variation on three-dimensional time-of-flight MR angiograms. Radiology. 1998 Apr;207(1):103-11.

19. Macchi C, Lova RM, Miniati B, Gulisano M, Pratesi C, Conti AA, Gensini GF. The circle of Willis in healthy older persons. J Cardiovasc Surg (Torino). 2002 Dec;43(6):887-90.

20. Urbanski PP, Lenos A, Blume JC, Ziegler V, Griewing B, Schmitt R, Diegeler A, Dinkel M. Does anatomical completeness of the circle of Willis correlate with sufficient cross-perfusion during unilateral cerebral perfusion? Eur J Cardiothorac Surg. 2008 Mar;33(3):402-8.

21. Fahy GM, Saur J, Williams RJ. Physical problems with the vitrification of large biological systems. Cryobiology. 1990 Oct;27(5):492-510.

22. Rabin Y, Plitz J. Thermal expansion of blood vessels and muscle specimens permeated with DMSO, DP6, and VS55 at cryogenic temperatures. Ann Biomed Eng. 2005 Sep;33(9):1213-28.

23. Weik M, Ravelli RB, Silman I, Sussman JL, Gros P, Kroon J. Specific protein dynamics near the solvent glass transition assayed by radiation-induced structural changes. Protein Sci. 2001 Oct;10(10):1953-61.

24. Meents A, Gutmann S, Wagner A, Schulze-Briese C. Origin and temperature dependence of radiation damage in biological samples at cryogenic temperatures. Proc Natl Acad Sci U S A. 2010 Jan 19;107(3):1094-9.

25. Glenister PH, Whittingham DG, Lyon MF. Further studies on the effect of radiation during the storage of frozen 8-cell mouse embryos at -196 degrees C. J Reprod Fertil. 1984 Jan;70(1):229-34.

26. Merkle, RC, Freitas, R,Jr., A Cryopreservation Revival Scenario Using Molecular Nanotechnology, CRYONICS, 29(4):7-8 (2008).

27. Gao W, Sattayasamitsathit S, Manesh KM, Weihs D, Wang J. Magnetically powered flexible metal nanowire motors. J Am Chem Soc. 2010 Oct 20;132(41):14403-5.

Is Cryonics Realistic?

Peter Gouras

Unrealistic but Rational

If we could quickly stop all the movement and action of every molecule in our body we could render it timeless. If we could reverse these changes we could reestablish a living state that had not aged in the interim. At present this is impossible but it is the dream of Cryonics that this can be done by using changes in temperature (1). But at present our temperature changes are relatively slow and non-uniform, which leads to problems. This causes the crystallization of water into ice and to erratic osmotic imbalances that regulate cell and cytoplasmic membranes. To counter this we try to use cryoprotective substances that prevent ice formation but at the concentrations needed they become in themselves toxic. Our inability to solve these problems makes Cryonics seem unrealistic to the scientific community and most probably to the world at large. What should we do? The first consideration is that our reasoning is rational. We do not believe in magic but only in scientific evidence. Rationality is our strongest virtue. But we must do more and this should be done as quickly and rigorously as possible. Is it worth pursuing this problem? From my perspective it is one of the most interesting and revolutionary problem we can investigate and therefore worth everything we can do to solve it.

Disbelievers

But this pursuing such a goal is not shared by most, even in the scientific community. Why? I suspect it is due to factors, probably genetic, that give people their world view. Each of us is slightly, sometimes enormously, different. Some are quick, others slow, some mathematical, others experimental, some religious, others agnostic or atheistic, some honest, others criminal, some conventional others imaginative or even crazy. There are vast differences in the makeup of mankind which are governed mainly by the neural circuitry in our brain and are difficult, almost impossible, to change by environmental factors. It is these neural circuits that govern the way we look at life and man's position in it. Our genetic heritage makes some enthusiastic cryonocists and others rabid opponents.

Darwinian Evolution and Death

The possibility of conquering death is a fascinating challenge to man's ingenuity worth the effort we can devote to it. It pits man against a cruel nature that discards us extravagantly and painfully by aging and death in its attempts to find the fittest. Once found, nature starts again in its deadly game of Darwinian evolution. In such a scenario, death is useful to clear the field for novelty and change that depends only on chance. We have now come to understand organic evolution. We can decipher the genetic script of every organism that has survived and even some that are extinct. Not only can we decipher the genetic code of any organism but we can also alter this code, changing the genetic makeup of organisms, even ourselves, by gene transfer. We can use such genetic engineering to cure babies, children even adults of horrid genetic diseases by restoring normal genes or blocking the action of deleterious ones. This is just the beginning of our ability to tailor our genome and not to rely on random mutations to blindly change things. This is an extraordinary advance that tempts our curiosity to follow its progress in the future.

Aging

The causes and eventual treatments of aging are being intensively researched. Scientists can now double the life span of small animals like mice and rats by caloric restriction. We are determining the gene expressions that caloric restriction activates in order to find out how to turn them on or off and thereby extend our life span without caloric restriction. There is no limit to what we can do in understanding the factors that cause us to degenerate and die within a small fraction of time relative to astronomical timescales. Aging is being investigated at the genetic and molecular level. Oxidative stress is a strong contender for producing many of the deleterious effects of aging. The oxidative powerhouse in all our cells, the mitochondria, cannot repair their DNA as nuclear DNA can and therefore accumulate mutations with time. This and the oxidative damage they produce is an important factor in aging but not the only factor. Other factors are involved which cannot be expanded here. By understanding why our cells age, we shall discover ways to minimize and eliminate aging in the future. There is an extraordinary world that could await us in the future if we can get there.

Past Progress

The pioneering studies of Audrey Smith and associates (2) succeeded in cooling hamsters, which are hibernators, to -2 C and rewarming them by diathermy to achieve healthy survival in about 50% of the hamsters. The Popovics (3) succeeded in lowering the body temperature of rats, which are not hibernators, to -1^{0}C for one hour in a super-cooled state (i.e. their body water still liquid) before thawing and reviving them. Most interesting is the approach used by Andjus and associates (4) in which they trained rats before cooling to just above freezing where heart-beat and brain waves ceased. Re-warmed by diathermy, the rats remembered their training tasks. Suda, a Japanese physiologist, reported that a cat cooled to below the freezing point of water recovered some neural activity after rewarming but this has never been followed up or confirmed. These experiments were done half a century ago and these laboratories are no longer active.

Present Progress

Currently two laboratories are doing cryo-preservation research. One is led by a distinguished pioneer in cryobiology, Greg Fahy, well known for his research on vitrifying kidneys. Fahy collaborates with a physicist, Brian Wowk, good pairing of expertise for this complex biophysical problem. They are supported by 21st Century Medicine located in Fontana California and financed by Saul Kent, a longtime promoter of Cryonics. Fahy and Wowk have made significant progress in reducing cryoprotectant toxicity, nucleation, crystal growth, and chilling injury. They show that an 8.4 M solution (VMP) designed to prevent chilling injury at −22 °C is nontoxic to rabbit kidneys when perfused at −3 °C. This permits perfusion–cooling to −22 °C with only mild damage. They demonstrate that the kidney tolerates a 9.3 M solution known as M22, which does not devitrify when warmed from below −150 °C at 1 °C/min. If M22 is added and removed at −22 °C, it is uniformly fatal, but when perfused for 25 min at −22 °C and washed out simultaneously with warming, postoperative renal function recovers fully. When kidneys loaded with M22 at −22 °C are further cooled to an intrarenal temperature of about −45 °C (halfway through the putative temperature zone of increasing vulnerability to chilling injury), all kidneys support life after transplantation and regain normal creatinine values. However, medullary, papillary, and pelvic biopsies taken from kidneys perfused with M22 for 25 min at −22 °C were found to devitrify when vitrified and re-warmed at 20 °C/min in a differential scanning calorimeter. It is not yet known whether this devitrification is seriously damaging or can be prevented by improving cryoprotectant distribution to more weakly perfused regions of the kidney or by re-warming at higher rates. This very important research implies that success with cryopreservation will come from careful control of multiple factors such as perfusion rate and pressure, the effectiveness and toxicity of the cryoprotective agents and the rates of cooling and re-warming. This success requires rapid treatment of the excised organ. In the cryoprotective treatment of humans this immediate opportunity to start perfusion and cooling is seldom achievable. The second laboratory currently involved in cryopreservation research addresses this difficulty that occurs with most human cases. This laboratory involve is run by Aschwin and G Dewinn located in Oregon and supported in part by Ben Best. They are trying to understanding how to deal with the No Reflow phenomenon, which is a vascular resistance to flow, which develops after a period of ischemia. Cooling of the brain during these periods of no-flow tends to mitigate this handicap.

Future Progress

I believe that we should try to obtain financial support for several more cryobiology laboratories. I suggest that we furnish one such laboratory to examine the cooling tolerance of a small animal like a mouse or rat. There are several reasons that I recommend this course. First it would involve studying how the brain tolerates the same treatment that kidneys can tolerate. Second the use of an entire organism would give us a rapid feedback on what is success or failure. With kidney research one must transplant the organ back into the same animal that had it removed, remove the untouched kidney and then wait days or weeks to know whether a improvement in cryopreservation has occurred. Using an entire animal provides a rapid assessment of success or failure. Is there a heart-beat, a muscle movement or a pupil reaction to light after the animal has been re-warmed? These responses are quickly detectable providing rapid insight into the effects of the cryo-preservation. If we could succeed in lowering the depth of cooling that allows such a small animal to survive compared to what has been obtained a half a century ago by the Smith laboratory, it would be press worthy and an enormous advance for the concept that Cryonics publicizes.

How do we fund additional laboratories? In the USA, the Cryonics Institute has about 1000 members. If each donated $100 a year to such a laboratory it would have $100,000 support, perhaps enough for a start. If the members of CI would contribute $1,000 a year this would provide such a laboratory with 1 million dollars which could attract a competent cryobiologist. Whether similar support could be obtained for a laboratory in Europe is moot at present.

Cryonic's Future

Predicting the future is difficult because of the complexity, variety and competitive nature of human life and the unpredictability of natural disasters. The best way to speculate on the future is to examine the past. Historically man advanced enormously ever since genetically separating from his nearest primate ancestor the chimpanzee about 7 million years ago. These advances are due to the evolution of man's cerebral cortex permitting him to analyze and control nature. Most of this progress occurred over the past several hundred years that indicates acceleration in scientific progress. The manipulation of energy, flying, rocketry, computing, quantum and astrophysics have occurred mainly if not totally through systematic experimentation. Man has promoted medical research to optimize the quality and duration of our lives and progress continues. Infectious diseases are being eliminated even viral ones. Cancer is slowly being conquered. Aging is being investigated with the ultimate view of counteracting it. Death itself is now questioned by the Cryonics movement. The use of computer guided electronic chips as sensory and motor aids is growing. I am confident that the idea that Cryonics has awakened will continue to grow in the future although I doubt that anyone one of us will be revived in this century. But if we don't try we shall never know. What I fear most is the unpredictable. Many changes in man's behavior and environment will occur in this century, which are not easily predictable. This unpredictability is troublesome. The evolution of the computer and robotics will certainly change our future world and could become a threat to organic life.

References

Feinberg G. Physics and Life Prolongation. Physics Today November 1966, 45-8

Smith AU Biological Effects of Freezing and Super-cooling. Williams and Wilkins Baltimore, 1971

Popovic P, Popovic VP Cryobiology 2:23, 1965

Andjus RK, Knopfelmacher F, Russell RW, Smith AU Nature 176:1015. 1955

Suda I, Kito K, Adachi C. Viability of long term frozen cat brain in vitro.

Nature. 1966 Oct 15;212(5059):268-70.

Fahy GM Wowk, B, Wu J, Phan J, Rasch C, Chang A, et al. Cryopreservation of organs by vitrification:perspectives and recent advances. Cryobiology 2004; 48:157-78

Fahy GM, Wowk B. Pagotan R et al Physical and Biological Effects of Renal Vitrification. Organogenesis 5 (3): 167-75

Internal ethical issues in cryonics

Sebastian C. Sethe

Introduction

Exposition

The notion that the inevitability of aging and death may in some way be negotiable by technological means has attracted considerable ethical debate. The practice of cryonics – the suspension at cold temperatures of a body, with the hope that healing and resuscitation may be possible in the future – has sometimes been referred to in these discussions; however, focussed ethical analysis of the practice is rare. In the present analysis, I will not consider ethical considerations that cryonics shares with the wider debate on drastic life-extension (such as 'overpopulation'[1] and including the putative association with 'transhumanism' discussed elsewhere in this volume) and I try to avoid rehearsing aspects of the (by now well-explored) debate regarding the ethical implications in light of potential social, economic, philosophical and psychological consequences of life extension. Instead, this essay attempts a brief examination of the perhaps 'underreported' ethical issues that arise in cryonics as they are faced by a cryonics practitioner ('cryonicist'). I will call this the 'Analytic Perspective' for present purposes. After a brief exposition on what that perspective entails, followed by an example applying the Perspective to one of the central moral issues in cryonics, I will venture to chart various issues in advanced 'cryoethics' in a basic declination of moral theory: Sub-chapter II draws on Utilitarian and Communitarian approaches and ethical challenges to cryonics that

---

[1] Shaw D. "Cryoethics: seeking life after death" in: Bioethics. 2009 Nov;23(9):515-21.

may be advanced thereunder. We then move to ethics of the 'inner sphere' (III) including deontological, and virtue ethics approaches. Lastly, in IV, I will consider cryonics as subject for 'specialist' professional ethics in medicine and business. A very brief outlook concludes the analysis.

The analysis aims to chart issues in moral philosophy rather than to present an in depth analysis.

Analytic Perspective

This analysis relies on taking central account of the perspective of the cryonics practitioner (the Analytic Perspective- AP). An important consideration in adopting this perspective is that cryonics is conducted voluntarily by all parties and that practicing cryonics has very limited direct effect on others (exceptions and implications will be discussed below).

In adopting the AP it is important to grasp the central dilemma of cryonics: While the reasonableness of expecting cryonics 'to work' is certainly contested[2], cryonicists have concluded that as it is practiced, or as it feasibly could be practiced, with current technology, cryonics has a sufficiently solid scientific foundation to [3]/[4]/[5] to make the possibility of eventual 'success' not entirely implausible. On the other hand, all cryonicist have to agree that there is no empirical evidence that cryonics as practiced currently has ever 'worked' – no human has ever been brought back alive from cryonic

---

[2] Darwin, M "Cold War: The Conflict Between Cryonicists and Cryobiologists" in: Cryonics June 1991 pp.4-17 & July 1991 pp.2-15

[3] Best, B.P. "Scientific justification of cryonics practice" in: Rejuvenation Research (2008); 11 (2), pp. 493-503

[4] Whetstine, L. , Streat, S. , Darwin, M. "Pro/con ethics debate: When is dead really dead? " in: *Critical Care* Volume 9, Issue 6, December 2005, Pages 538-542

[5] Lemler, J. , Harris, S.B. , Platt, C. "The arrest of biological time as a bridge to engineered negligible senescence" in: Annals of the New York Academy of Sciences (2004) 1019, pp.559-563

suspension and there is universal agreement that such resuscitation is not possible with today's medical technology.

Therefore the Analytic Perspective reveals a dilemma of presenting two apparently contradictory characteristics: Cryonics, as it is practiced currently, is at once

(characteristic A) a potentially life-saving endeavour AND

(characteristic B) a potentially futile endeavour.

As I will demonstrate, almost all ethical reflections regarding cryonics need to take account of these two characteristics. They are the minimum and complete set of criteria. The Analytic Perspective does not seek to anticipate which moral theory or ontology a cryonics practitioner adopts.

Example: Moral status of the cryonically suspended

To illustrate the operation of the Analytic Perspective and to deal with another central consideration in the ethics of cryonics.

Central to the cryonics case is the argument regarding the arbitrariness of diagnosing death which has undoubtedly experienced historical shifts and continues to be subject to debate in the practice of medicine.[6] This creates a conceptual challenge of the moral status of the person who has been cryonically suspended. While cryonics requires the legal death of a person to proceed, to speak of a suspended person as 'dead' establishes assumptions that cryonics exists to rebut. This may seem more intuitively obvious where the cryonics suspension process has only just begun. (As Wowk comments: "Two minutes of cardiac arrest followed by restoration of blood circulation does not a skeleton make"[7]) and more difficult to conceptualise once the body has been fully suspended.

---

[6] Whetstine L., Streat S., Darwin M., Crippen D. "Pro/con ethics debate: When is dead really dead?" (2005) Critical Care, 9 (6), pp. 538-542.

[7] Wowk, B "Medical Time Travel" in: *The scientific conquest of death* (2004) Libros

A crude test to explore the strength of this dilemma is to present a variant of the infamous 'trolley problem'[8]: consider a runanway railway that threatens to collide with an occupied pram ... or could be diverted crashing into a cryonics facility destroying the dozens of bodies suspended therein – and various permutations of that scenario including, as a train driver, *having* to make a selection between both 'targets'.

From the Analytic Perspective, this is may be a real and painful ethical problem since interred bodies are not dissimilar in moral status from patients on life support. The answers (which we will resist exploring here) will reflect the degree to which the cryonicist emphasises the above outlined characteristic A over B. Ultimately, a cryonicist might decide that the difference between actual and potential life is decisive (such as a committed Christian may still elect to save the baby in the pram over thousands of frozen embryos for in vitro fertilisation) or that other factors (proximity of causation, kinship to a particular suspendee) may be most pertinent.

Whatever the decision, I merely aim to illustrate that in the context of the present discussion one would fail to appreciate the ethical foundation of the matter if one fails to apply the Analytic Perspective to it.

Utilitarianism & Communitarianism

Money well spend?

It has been suggested that cryonics is ethically dubious because it is a substantial investment into speculatively prolonging a single life, whereas the same money could be spend more effectively on prolonging a great number of lives with greater surety.

---

[8] Thomson, Judith Jarvis "The Trolley Problem" in: 94 Yale Law Journal 1395-1415 (1985)

The costs of cryonics vary considerably, ranging between $40.000–$200.000 in the aggregate.[9] According to some estimates, this amount could save between 49-244 lives in the poor world (based on a figure of $820 per life)[10]. Such calculations are certainly fraught and their applicability dubious, but it remains the case that on a strict utilitarian calculus much of the lifestyle of the affluent world is ethically problematic and distributive measures would appear an ethical imperative.[11] Consequently, substantial investment into purely personal projects could be considered immoral – whether it is buying a house, a holiday or a cryonics contract. A number of objections are plausible. Firstly a cryonicist might reject utilitarian reasoning altogether, which would take us outside of the remit of this section. However, it should be remarked that the 'rationality' of consequentialism would appeal to many cryonicists not least because many cryonicists do not derive their ethics from religious stricture.

Even a cryonicist who operates a strong utilitarian moral theory could reject utilitarian reasoning in the personal case – this is because for the cryonicist, the cryonics endeavour is not a 'luxury' but a self-preservation measure. Even stern utilitarian regimes typically fall short of compelling the individual to sacrifice their own life for the greater good.[12]/[13] However, in line with the first principle, the cryonicist cannot be assured of the self-preserving effectiveness of cryopreservation.

---

[9] Rough estimate, based on a median of costs from KrioRus, Cryonics Institute, Alcor taking account of associated costs such as standby and transport.

[10] based on givewell.org and Peter Singer "The Life You Can Save" Random House 2009

[11] Barry, C., Valentini, L. "Egalitarian challenges to global egalitarianism: A critique" in: Review of International Studies, 35 (2009) (3), pp. 485-512.

[12] Draper, K.; "The personal and impersonal dimensions of benevolence" 2002Nous 36 (2), pp. 201-227

[13] Overvold, M. C. (1980). 'Self-Interest and the Concept of Self-Sacrifice,' Canadian Journal of Philosophy 10. pp. 105–118.

Nonetheless, a cryonicist might mobilise some of the following counter arguments from within a utilitarian mindset:

   a) "Others invest in poverty reduction. By making that financial contribution I strengthen the fledgling enterprise of cryonics provision, which I consider a moral good, which may not otherwise exist"

   b) "I purchase the inner peace that I require to lead an ethical life"

   c) "If cryonics succeeds, I will be able to do more good in the long run".

These arguments are illustrative only. It can certainly be observed that consequentialists have not historically agreed on the equal distribution problem.[14] In any event, we cannot assert that from a utilitarian perspective cryonics is inherently more unethical than other lifestyle investments. It also is worth mentioning that cryonics is typically paid for by carefully accumulated savings. Even utilitarian theorists tend to be more tolerant of singular 'folly' that an individual has worked long and hard for to obtain.[15]

Communitarianism

Are you a burden on future generations?

Given that the modalities of future revival are fraught with utter uncertainty, I would leave aside entirely the field of 'revival ethics' – ethical considerations that apply when attempting to revive a frozen 'patient'. However, even the storage of a patient is an ongoing burden on somebody. Currently, cryonics facilities are run by essentially the 'second generation' of cryonicists – those who may still have met some of the first suspendees but might have been

[14] Weinstein, David (2007) Utilitarianism and the New Liberalism; Cambridge University Press
[15] Miller, Richard. 2004. 'Beneficence, Duty and Distance', Philosophy and Public Affairs 32:357–83

quite young at the time. The current business model of both established providers stipulates that cryonics storage is a perpetual service but a charge is only levied once. This may mean that, at least
over time, the provider experiences the responsibility of watching
over a non-paying customer as a burden. (Interestingly, a situation
where a smaller sum was provided upfront and a 'cryonics rent' be
paid out by relatives is seen as much less realistic or desirable.) As
an analogy, in some contexts graves are only maintained as long as
the space is paid for, in others one waits for the stone to wither. In
this case capitalism rides to the rescue: the cryonicist has 'paid' to
be maintained. She does not need to feel that she is imposing an
immediate burden. If future generations consider this payment no
longer sufficient, they can relatively easily abandon the maintenance of the suspension since the cryonicist is in no position to
challenge this. Thus, if one is concerned about 'being a burden' it
would seem advisable to not establish very stringent legal protec-
tions (although, if one is demonstrably a burden it seems unlikely
that future courts would or could enforce continued suspension).

Are you contributing to social inequity?

Cryonics is expensive. Less expensive than some seem to assume
and through insurance funding within reach of the broader middle
class of affluent countries, but certainly not affordable by all who
want it. It could be argued that therefore a cryonicist is exacerbat-
ing social inequality. This point however does not stand in current
circumstances. Counterarguments could be that high prices for early adopters[16] could decline as cryonics becomes mainstream, or
simply that "we do not normally think it an ethical requirement to
prevent good being done to some unless and until it can be done to
all"[17] – however both would miss the reality of cryonics: despite

---

[16] Platt, Charles; "Hamburger Helpers" in: *Cryonics*; 4th 1998, pg.13-16
[17] Harris, J. (2002). "Response to Glannon"; *Bioethics*; Vol.16, No.3, pp.284-291.

being available for decades, the practice is not widely seen as 'good' and consequently has not been widely 'adopted'. Given the choice, most seem to prefer a free holiday to a free cryonic suspension contract.[18] Whether those who go on holiday are immorally contributing to social inequality is another matter, but the widespread acceptance of the practice allows us to dismiss the allegation of moral misconduct at least on the wider social front and at least for the time being. However, for a committed cryonicist the value of a cryonics arrangement is of vastly greater value than a holiday – it is the potential difference between life and death. Therefore, inequity remains a moral problem in the small sphere of those who yearn for cryonics. This is mainly a problem for those who cannot get affordable life insurance i.e. the severely disabled or citizens of poor countries. It is unclear how high these numbers are. Cases are known where cryonicists have supported the underprivileged through donations[19]. Given the size of the community, these ad-hoc efforts at charity may currently be the best solution if inequitable access were to be perceived as a serious moral challenge. This may require a more open and structured approach by the cryonics community to support individuals who desire cryonics but for whom it is firmly out of reach.

Kinship

Whether acknowledged or not, some ethical positions accord special respect and importance to kinship and companionship. The following considerations are example challenges, focussing on how a choice for cryonics affects the cryonicists' closest social sphere.

---

[18] Lawrence Wilson, Juliet "A cool prize that is just too hot for me to handle" in: Edinburgh Evening News; 25 September 2002 (http://living.scots man.com/julietlawrencewilson/JULIET-LAWRENCE-WILSON-A-cool.2 363906.jp)

[19] http://www.longecity.org/forum/topic/27078-immortality-institute-orights-matching-fund/

"Are you pressuring loved ones to confront situations that they find difficult or abhorrent?"

As an example, the cooperation of the next of kin can be very important. However, these people may consider the concept or the practical details of cryonics difficult to contemplate or engage with. As we can see from other examples where 'lifestyle' choices can offend others, (e.g. disclosing a same sex relationship to parents/grandparents), these are very individual constellations depending on who and how the problematic content is broadcast. A cryonicist following pragmatic ethics would need to consider how to minimise offense to others. Luckily, in a pluralistic society and considering that in some places cryonicists are numerous enough to offer mutual assistance, such consideration should come relatively easily.

"Are you pressuring loved ones into joining?"

From many a cryonicist perspective, those who are not receiving cryonics services waste a chance of a 'second life', potentially in their company. The 'temptation' might therefore be great to pressure a loved one into participating in cryonics. Parallels to religion may come to mind. Given the uncertainty principle, pressuring someone into cryonics against their better judgement would not seem an ethically desirable outcome (although from the perspective of personal/virtue ethics, a cryonicist who does not at least attempt to gently persuade a loved one of the merits of cryonics may not be acting ethically either).

Personalised Ethics

Dignity and the Golden Rule

The 'father of cryonics' Rober *Ettinger* has reflected on the applicability of deontological reasoning to "The Golden Rule should work better for immortals than for humans, even as a mere tactic of expediency".[20]

Scott *Stroud*[21] uses Kant's infamous statement on "without a God and without a world invisible to us now but hoped for, the glorious ideas of morality are indeed objects of approval and admiration, but not springs of purpose and action" as vantage point to critique Ettinger's position "which brings with it a conception of morality that does not grasp the insights resident in Kant's position – the quest for holiness of will and commensurate happiness is only available in a hoped for moral realm, not in the world of sense." One ventures here into the query of 'immortality' as a metaphysical aspiration.

If a cryonics arrangement is established and conceptually endorsed as 'the only way' of dealing with death, the cryonicist is faced with a conceptual dilemma when those die who do not benefit from a cryonics suspension: there is, categorically, no hope for them. This may seem a trivial point; it is after all the dilemma faced by all atheists. However it could be argued that the pursuit of cryonics renders the framing of death as hopeless more imminent.

Also arguing from a Kantian tradition, *Beyleveld* suggests that "trying to live forever", is not only futile it is also morally flawed: "The mere aspiration is a refusal to accept conditions that are necessary to give point to morality. As such it is an attempt to transcend moral responsibility. It is an attempt to transcend dignity by pretending that it does not exist. [...] this is the essence of undignified conduct, as well as being dispositionally dangerous."[22] In contrast, aspiring to live an undetermined amount of time (while not free from other problems) is not seen by *Beyleveld* & Co-author

---

[20] Ettinger, Robert, *Man into Superman*; St.Martins Press; 1972

[21] Stroud, Scott "A Kantian Critique of Cryonic Immortality" in: Charles Tandy & Scott R. Stroud (eds.), *The Philosophy of Robert Ettinger*, 2002, pp.135

[22] Deryck Beyleveld, Roger Brownsword; *Human Dignity in Bioethics and Biolaw*; 2001; Oxford University Press

Brownsword as essentially problematic: "Because such extension recognises the vulnerability of bodily existence...". Interestingly, an aspiration to 'immortality' is not endorsed by all thought leaders in cryonics. *Best* advances strongly that talk of such concepts is unwarranted in cryonics.[23] If a cryonicist thus conceptualises the endeavour as a 'life extension' project, both the allegation of 'undignified conduct' and of dwelling in the 'sensuously conditioned' does not stand as inevitably connected. Moreover, it bears revisiting what it might entail, from a deontological perspective' to rely on cryonics as 'springs of purpose and action'. Certainly –to place it in the familiar frame– cryonics is seen by its practitioners as 'a means to an end, not as an end in itself'. This end one presumes, is neither to be frozen, nor indeed 'immortality' – but instead to live one's life as a basis for 'dignified conduct'.

Virtue Ethics

From a virtue ethics perspective, the cryonicist needs to consider how her preference and engagement with cryonics reflects ethically on their own character. Very different and competing frameworks for analysing virtue ethics positions exist.

Question that a cryonicist may ask include.

"Are you ruled by fear?"

Cryonics could be seen as infringing the classic virtue of 'Fortitude' (andreia), by creating a nexus of relationships, thoughts and actions in the cryonicist's life that are motivated by fear. In some sense, this might be an observable phenomenon – are cryonicist more timid because of their focus on managing death? I find no evidence for such an assumption. There is some empirical indication that those who deal with death in an ordered manner become less anxious

---

[23] Best, Ben "Some Problems with Immortalism" in: BJ.Klein, S.Sethe et al. (eds) The Scientific Conquest of Death – Essays on Infinite Lifespans; 2004; Libros En Red; pp.233-238

about it.[24] Anecdotally, I have met cryonicists who seemed quite constrained in their capacity to experience life positively by their fear of death and their need for constant precaution, and others who had a very carefree and adventurous attitude. Neither observation is generalisable, nor would it shed light on the virtuous courage of the person. This is the case not least because 'Fortitude' may well manifest itself in other forms than in necessary link with mortal danger (such as in 'moral courage' although Aristotle himself might disagree[25]) and because 'Prudence' is in itself regarded as a key virtue.

"Are you lacking in temperance?"

Cryonics could be conceived as a failure to uphold 'Temperance' (sophrosyne). This can be viewed as greed – cryonics has been described as a 'uniquely American strategy', the ultimate in neoliberal consumerism[26]/[27]. Whatever one makes of this analysis, the underlying allegation sees cryonics as an expression of hubris,[28] by aspiring to 'more than you are entitled' and breaking with the 'natural order of things' (infringing the roman virtue of pietas). This is a wider theme that cryonics shares with the discourse on 'posthumanism' and not one we can explore here. Suffice it to say that, in the context of virtue ethics, any concept of 'naturalness' is only meaningful if it appears as thus to the moral agent. Rather than dwelling

---

[24] Harrawood, L.K. , White, L.J. , Benshoff, J.J. "Death anxiety in a national sample of United States funeral directors and its relationship with death exposure, age, and sex" in: Omega: Journal of Death and Dying (2008) 58 (2), pp. 129-146.

[25] Cordner, C "Aristotelian Virtue and Its Limitations" in: Philosophy (1994), 69: 291-316

[26] Krüger O; "The Suspension of Death. The Cryonic Utopia in the Context of the U.S. Funeral Culture" in: Marburg Journal of Religion: Volume 15 (2010)

[27] Romain T "Extreme Life Extension: Investing in Cryonics for the Long, Long Term" in: Medical Anthropology, (2010) 29:2, 194-215

[28] Iserson, KV. "From Creatures to Corpsicles: Man's Search for Immortality" in: HEC FORUM Volume 16, Number 3, 160-172

on concepts of greed and hubris, it may be useful to return to the 'traditional' virtue of Temperance as avoiding excess. This raises the question in what way the cryonicist is behaving excessively. If the allegation is one of seeking an 'excess of life' such an allegation would not seem easy to maintain: in the context of virtue ethics life itself has no quality – the question is whether one leads a 'good life' and to answer the question of whether there is an intrinsic temporal limit to 'eudaimonia' simply by pointing to established life expectancy seems a rather lazy manoeuvre.

Spirituality & Selfhood

This analysis focuses on moral philosophy – consideration of morality based on various spiritual and theological considerations would transcend the present scope. Briefly, while a 'rationalist' or 'materialist' viewpoint may fit particularly well with interest in cryonics[29] a range of sources have argued that religion and cryonics are not incompatible[30]/[31],

Another question of wider ethical significance is whether cryonics can or could ever achieve what it sets out to do in a philosophical sense: has the 'resuscitated' cryonicist succeeded in the quest to extend one's life, or is the person thus reconstituted 'someone else'? Is cryonics necessarily a "self-defeating fantasy"?

A significant amount of thought – not least among cryonicists themselves – has been extended on this topic.[32] It may depend on the technical mode of resuscitation (e.g. is one reviving the frozen body itself, or a bio-or nanotechnologically constructed viable

---

[29] Perry, R. Michael; Forever for All: Moral Philosophy, Cryonics, and the Scientific Prospects for Immortality (2000);

[30] Montgomery, John Warwick; "Cryonics and Orthodoxy" in: Christianity Today, 12, 816 (May 10, 1968)

[31] Ryan, Derek; "Cryonics and Religion" in: Cryonics, Volume 11(12) Issue 125 (December 1990) pp.21-23

[32] Hughes, J. "The Future of Death: Cryonics and the Telos of Liberal Individualism" in: Journal Of Evolution And Technology, Volume 6, July 2001

'copy', or a faithful digital simulation at the subjects brain processes at molecular level...). Much of this debate requires setting out a theory of personhood, which is clearly beyond the scope of this article. One can sidestep the debate in asserting the following: from the Analytic Perspective, the objective is some continuity of self. If the cryonicist were to find arguments compelling on a moral level, which stated that such is impossible, the attempt would not even be made. Eventually, the question may turn, for many practical ethical purposes, on the self-assessment of the revivee.

Professional Ethics

Law

The following section considers 'professional ethics' as a set of informal (and sometimes contested) strictures. Focussing on ethics, I will leave aside specific questions of cryonics and the (codified) law. Legal issues do of course arise at these junctures and most have changed little in half a century [33]/[34]/[35].

Of course, where I have touched on points where cryoethics and the law are at odds, I in no way mean to infer that current cryonics practice is or would ever contemplate breaking the law.

However, the question of whether the law 'does right' by cryonicists, indeed whether from the perspective of the cryonicist there is an obligation to observe a law that may lead to a morally undesirable outcome is not settled. The question of civil disobedience is of

---

[33] Henderson, Curtis; Ettinger, Robert C. W. "Cryonic Suspension and the Law" in: 15 UCLA L. Rev. 414 (1967-1968)

[34] LaBouff, John Pau "He Wants To Do What - Cryonics: Issues in Questionable Medicine and Self-Determination" in: 8 Santa Clara Computer & High Tech. L. J. 469 (1992)

[35] Sullivan, Ryan "Pre-Mortem Cryopreservation: Recognizing a Patient's Right to Die in Order to Live" in: (2010 / 2011) 14 Quinnipiac Health L.J. 49

course a complex debate in healthcare jurisprudence and beyond[36] – from the Analytic Perspective it will again depend on weighting Characteristics A and B, as well as the contingent variables of the specific case. Pragmatically, a committed cryonicist needs to consider the effect legal noncompliance may have on the whole fledgling enterprise. In fact, given the small size of the field und its proclivity to attract sensationalist media coverage, the same power to, by misconduct or carelessness, spark a reaction that could complicate or end entirely all or much of cryonics practice accrues to many of those who assume an active role in cryonics, a situation that brings its own ethical constraints.

Medical Ethics

Cryonics is cast as a medical undertaking – cryonics providers refer to 'patients'. Indeed the first part element of the Analytic Perspective would suggest that Cryonics may at some point be the only potentially life saving treatment available. From this perspective, withholding such treatment would be ethically untenable. However, the Analytic Perspective in its second part recognises the uncertainty element. What does this mean regarding the moral claim that a cryonicist has on a physician? Issues arise mainly if the medical practitioner is unwilling to assist the cryonics practitioner.

Non-cooperation

Although in practice, these attitudes will almost always be joined, a distinction can be drawn in theory regarding the physician who refuses because she considers it (1) futile or (2) morally repugnant for other reasons.

Regarding (2) it is difficult to make a moral claim on a practitioner that forces action against moral conviction. It is established

---

[36] Childress, James F. "Civil Disobedience, Conscientious Objection, and Evasive Noncompliance: A Framework for the Analysis and Assessment of Illegal Actions in Health Care" in: J Med Philos (1985) 10 (1): 63-84.

that this op-out has limits roughly prescribed by social consensus, often also iterated in law. Three tiers of abstention are recognised: full abstention that may leave a patient unable to obtain the treatment (a surgeon may refuse to practice a particular cosmetic surgery); abstention where another colleague is likely to perform the service (a Christian physician may refuse to perform an abortion but the service needs to be available where it is legal); and refusal of abstention (a white supremacist refusing treatment to a black person would be guilty of misconduct). For cryonics, we can at least observe that there is no social consensus regarding the acceptability of the practice. The fact that it is legal in many jurisdictions may serve as a guide that outright obstructionism or may not be morally justified, but beyond that, a cryonicist may have a very weak claim on the non-cooperative physician motivated by repugnance.

However, if the main grounds for refusal are that the physician is simply unable to see any merit in the practice, the real challenge is one of empathy – can the physician recognise the earnest desire of the subject for receiving cryonics.

Non-maleficience

Fundamental difficulties arise where cryonics practice establishes a demand to perform in a manner contrary to established medical ethics, in particular where measures to facilitate cryonics could hasten 'death'. Cryonics begins essentially after legal death has been declared. There is do doubt that for the cryonicist this is often a purely bureaucratic barrier that stands in the way of optimal cryonics practice and that from purely technical perspective a more controlled process of transitioning from life to cryonic suspension would greatly minimise the putative barriers to later resuscitation (Interestingly, often one of the first medications administered to the subject is a sedative to prevent their 'reawakening' so that the cryonics suspension may proceed unimpeded).[37] In many cases, this apparently acute ethical conundrum could be eased by putting it into the proper context of end-of-life care: While not uncontested, it is widely accepted in medical ethics that a physician may apply measures that ease the suffering of the moribund patient even if they hasten the death that is near anyway. For a committed cryonicist, initiation of cryonics procedures in a controlled manner may significantly ease existential dread. From this perspective, cryonics as an element of compassionate palliative care might seem feasible.

But one is of course on thin ice in pursuing this line of argument: Obviously, no physician in favour of a controlled legalisation and medicalisation of a 'right to die' could be opposed to exercising this right in order to facilitate cryonics (although there may still be significant difference between fledgling schemes of physician assisted suicide and regimes that facilitate cryonics which may or may not need to involve a greater degree of active participation by others), but the topic becomes more complex where death is not imminent: A cryonicist suffering from a neurodegenerative condition may seek early suspension because at the time when 'natural' death would be pronounced cryonics of the strongly degenerated brain may no longer be considered worthwhile.

Cryonics as a medical experiment

These and other challenges illustrate how cryonics can complicate the practice of medicine, but medical ethics have become used to incorporating novel technologies into a continuing discourse. Clearly, cryonics cannot currently be considered a staple consideration in medicine and although cryonicists may feel disadvantaged by this fact, the Analytic Perspective, resting as it does on uncertainty, could not easily compel a shift in ethically motivated medical decision making. However, there is a dimension of medical ethics that

---

[37] E.g. "Alcor Patient A-1705" in: Cryonics; Vol. 22:2; 2nd Quarter 2001

may find more effective ways of engaging with cryonics: the potential medical affiliation of cryonics is not one of established medical practice and more akin to a high-stakes clinical trial.[38]

Business Ethics

From the Analytic Perspective, cryonics is mainly a medical undertaking, while in practice it is enabled and funded through commercial transactions. Even in established fields medicine and business do not make for easy bedfellows (although this tenet is not uncontested[39]).

Still, there are general ethical challenges that face cryonics providers, as part of 'standard' business ethics: Transparent and fair dealing, accountability and sustainability, avoidance of cronyism and conflicts of interest, upholding quality and privacy, fair competition and pricing. Among these issues, the most obvious might be the duty to advertise fairly.

As other businesses, cryonics providers are also challenged to a wider sense of stakeholdership beyond their customers. There are ethical responsibilities vis a vis employees, collaborators, volunteers, communities and the environment. Dealing sensitively with the relatives and companions may be the most salient among these, since providers are by necessity placed at the emotive transition of life to legal death and beyond.

Thus, from the Analytic Perspective, cryonics touches on analogous fields of ethically demanding scope including palliative care and funeral services without wishing to affiliate too closely with either. Interestingly, it has been observed quite independently that either field presents specific challenges to business ethics: "In some

---

[38] Merkle, Ralph "The Molecular Repair of the Brain" in: Cryonics magazine, (1994) 15(1/2).

[39] Gilmartin, Mattia J.; Freeman, R. Edward "Business Ethics and Health Care: A Stakeholder Perspective" in: Health Care Management Review April 2002 - Volume 27 - Issue 2 - pp 52-65

industries there is a natural lack of rational restraint on the part of consumers of which those out for maximum profits can take advantage, for example in the funeral or health-care industries"[40]. Indeed from the Analytic Perspective the existential importance of access to cryonics is on par with access to healthcare or funeral provision. While most countries have established regulatory supervision and compassionate provision regimes in these areas, no such support is of course provided for the out-of-pocket cryonicist. A number of cryonicists adopt a politically libertarian attitude, so it is doubtful whether institutional support and price constraints would be welcome.

This places the ethical responsibility of managing the 'lack of rational restraint' on the provider. In theory, market mechanisms should be effective, since unwarrantedly high profit margins on essential services (under the Analytic Perspective) should attract effective competition on pricing. In practice, the provision of cryonics services is such a niche enterprise that competition may not be forthcoming.

This uniqueness of cryonics provision exacerbates the ethical challenges to the cryonics provider. Other factors are the extreme technicality of the service, incorporating complex medical and scientific detail. Cryonics, as a practice, can be done more or less 'well', but what a technically 'good' cryonics service entails is difficult to assess. Technical details of what constitutes a 'good' cryonics suspension are difficult to grasp for a lay person and those specifics are themselves potentially contested among experts and in flux. There is no way for the suspendee to remonstrate for lack of diligence and in most cases even subjects' relatives and companions will not be able to seek effective redress after the fact (e.g. if a mistake was made during the subjects suspension, it may have caused

---

[40] Goldman, Alan H. "Business Ethics: Profits, Utilities, and Moral Rights" in: Philosophy & Public Affairs; Vol. 9, No. 3 (Spring, 1980), pp. 260-286

irreversible damage as measured against expectation of successful revival.)

Moreover, from the Analytic Perspective, the cryonics provider acts a custodian, entrusted with making existential decisions on behalf of the incapacitated suspendee and to do so potentially for many generations. While it is not unheard of for businesses to exist for hundreds of years[41], this will only be possible through careful management of assets and continued revenue flow. If the business acts imprudently the hopes and efforts of all preceding generations could be thwarted. Thus a cryonics provider is in the unique ethical position of having to justify its activities vis a vis a growing (but personally silent) stakeholder base. This is balanced against the ethical pressure to take risks in making cryonics available to as many aspirants as possible. This ethical precariousness can only be managed from a deep understanding of what has been the 'Analytic Perspective' herein. It would seem advisable for a cryonicist to seek out a provider whose key decision makers share this perspective.

In conclusion

In this analysis we have considered the practice of cryonics using a portfolio of moral theories and traditions charting a rather complex landscape of ethical challenges, some obvious, some more obscure, some easily dismissed some gaining in saliency only at second glance. Like an ambitious tourist guide, I have sought to avoid some of the more obvious places seeking instead to acquaint the newcomer with some of the more obscure yet worthwhile alleyways.

So what are the most *important* ethical issues in cryonics? According to this analysis, the most pertinent moral challenges, present themselves not to society and not even to cryonics practitioners directly but to those who provide cryonics.

---

[41] http://en.wikipedia.org/wiki/List_of_oldest_companies

While it may be very unwise for a bioethicist to downplay the ethical challenges of any subject, especially where one has been granted privileged access, I may need to disappoint on this front: Cryonics has many ethical challenges, some of them unique, some uniquely vexing – but, generally, the majority of ethical dilemmas that cryonics throws up are neither insoluble nor of global importance.

Cryonics remains a niche subject. Its practitioners are few and have not grown in number very significantly during the decades that the practice has been available. Generally, it is conducted by well informed consenting adults, its capacity to affect the lives of others outside of the friends, family and healthcare personnel of the cryonicist is very limited, it has very few environmental risk and its economic and political impact so far has been negligible. Although it receives a comparatively significant amount of media exposure (compared say, to orphan diseases afflicting the same number of persons as are committed to cryostasis), society at large has remained indifferent.

This is not to belittle the passions of those involved in cryonics, the very salient ethical challenges that can arise for individuals and its capacity –albeit so far unrealised– for effecting widespread social and moral debate and even change; however, I sought to demonstrate that if cryonics is to be taken seriously as a subject for moral analysis either on its own, as a case study or vantage point for wider reflections, the only way to do the topic justice is by engaging with what has been presented as the 'Analytic Perspective' in this article: the vantage point of one to whom the moral debate has pertinence. Only from this perspective can one hope to understand how cryonics will continue to present crushing moral dilemmas to some of its protagonists and retain the capacity to engender the greatest depth of existential despair and the most inspiring examples of human compassion, aspiration and flourishing. Moreover, I have aimed to show that by adopting such a perspective, however transiently, one is able to transcend the tired old vestiges of 'abstract' bioethical debate and engage on what I would not hesitate to characterise as an advanced level of moral discourse. This methodology in turn may open new perspectives not just on the field in question but on wider horizons of ethical theory and practice

Cryonics and Transhumanism – a Troubled Marriage?

Torsten Nahm

§1. Introduction

Regarded on a global scale, the cryonics movement is vanishingly small: Only around 3,000 people are members of a cryonics organization, corresponding to around 0.00004% of the population.

Things change dramatically when we look at the non-representative subsample of the population that makes up the transhumanist movement (which will be discussed in detail later). Of the members of the World Transhumanist Organization, 8% are also member of a cryonics organization, according to the latest published survey (WTA Survey 2007).

This means that the chances of being a cryonicist are around 200,000 higher than average if you are a transhumanist, an astonishing number indeed. When we look at the most prominent members of the transhumanist movement, the correlation seems even higher[1]. As we will show, this association extends not merely to a question of shared membership, but can be traced to common ideas and values that were present even in the earliest history of both movements.

Two questions immediately arise from these observations: First, why are cryonics and transhumanism so intimately linked? And second, is this symbiosis detrimental or beneficial for the cryonics movements?

---

[1] Ray Kurzweil, Max More, Eric Drexler and Marvin Minsky are among the most visible transhumanists, and are all signed up for cryonics.

§2. Discussion

Before we start our discussion of the relationship between cryonics and transhumanism, let us first take a detailed look at the transhumanist movement.

The transhumanist movement originated in the 1970s and concerns itself with the future evolution of humans and human society. Transhumanists believe that humans can and should use technology to overcome perceived human shortcomings and weaknesses, such as disease, aging, limited intelligence or moral deficiency. In the words of the Transhumanist Declaration, transhumanists "seek personal growth beyond our current biological limitations". Transhumanists believe that technology is the main driving force behind human progress, and see it as the means to take evolution into their own hands. Transhumanists are generally open to human enhancement, including genetic engineering and "cyborgization" of their own bodies.

Transhumanism is based on a strong core of humanist values, including self-fulfillment and personal freedom. For this reason, while transhumanists advocate the use of technology on their own bodies, they make it clear that any enhancement should be an informed, personal choice for an individual that does not cause harm or danger to others or society.

FM-2030 (born 1930 as Fereidoun M. Esfandiary) is considered by many as the founding father of transhumanism. He also contributed (in particular through his 1989 book, "Are You a Transhuman?") in giving a name to the nascent movement. In his "Upwingers Manifesto", published in 1973, he writes:

> "Specifically we want to marshall humanity's genius to overcome our supreme tragedies – aging and death. We want to help accelerate the colonization of our solar system and open up this infinite Universe of infinite space, infinite resources, infinite potentials."

In the last 10 years, transhumanism has moved forcefully onto the world media stage. In 2002, the Roco-Bainbridge report on "Converging Technologies for Improving Human Performance" espoused transhumanist ideas relating to intelligence enhancement, stating:

> "Thus, the engineering of mind is much more than the pursuit of scientific curiosity. It is more even than a monumental technological challenge. It is an opportunity to eradicate poverty and usher in a golden age for all humankind."

The report, co-sponsored by the National Science Foundation, received wide-spread media attention and lead to extensive debates on the possibility and desirability of advanced artificial intelligence and enhancement in general.

Transhumanist ideas, along with detailed analysis of the course of recent technological advance, were further popularized in the book "The Singularity is Near: When Humans Transcend Biology" by Ray Kurzweil in 2005. Kurzweil claims that technological progress advances exponentially and predicts that by 2045 the so-called Singularity will occur, an extremely disruptive event that will completely change the face of the world. Kurzweil writes:

> "They key idea underlying the impending Singularity is that the pace of change of our human-created technology is accelerating and its powers are expanding at exponential pace."

He describes the Singularity as

> "a future period during which the pace of technological change will be so rapid, its impact so deep, that human life will be irreversibly transformed. Although neither utopian nor dystopian, this epoch will transform the concepts that we rely on to give meaning to our lifes, from our business models to the cycle of human life, including death itself."

However, for all the media attention in recent years, transhumanism (like cryonics) remains a rather small movement. The World

Transhumanist Association[2], the movement's world-wide umbrella organization, counts about 6,000 members.

We are now well placed to examine the shared history of cryonics and transhumanism and their intimate relationship. Indeed, this shared history can be traced to the earliest protagonists: Robert Ettinger, the founding father of cryonics, saw cryonics not simply as a groundbreaking medical technique, but as a necessary step in transcending human limits. His publication of "The Prospect of Immortality" in 1962 was followed in 1972 by his book "Man into Superman". In this, he asserts:

> "To the best of my knowledge this book is the first of its kind–the first to deal in a reasonably systematic way with the varieties and potential of superhumans. ... You personally and your families have a genuine opportunity to prolong your lives indefinitely and outgrow the human mold."

For Ettinger, simply becoming immortal was not enough. He felt a human life merely prolonged, but not enhanced or expanded, was not worth the effort. In the introduction to "Man into Superman" he writes:

> "Merely to expand time, without expanding the psyche, seems to hold little attraction."

We can thus see that for Ettinger, cryonics and transhumanism (although at that time not yet called by this name) were inextricably linked: for those with the ill fortune of an early death, cryonics is the technology by which they may yet become transhuman, and without transhumanism, cryonics would be futile. While gaining far less reception than "The Prospect of Immortality", "Man into Superman" establishes Ettinger as one of the earliest transhumanist thinkers.

---

[2] The organization is currently marketing itself by the name "Humanity+" in an effort to be more accessible

These ideas are closely echoed by FM-2030, whom we met as the father of the transhumanist movement above. FM-2030 was an immortalist and had a strong desire to live in his envisioned transhumanist future; indeed, he chose his name based on his intention to live to at least 2030. He was a vigorous advocate of cryonics: to him, anybody not signed up for cryonics was throwing away their chance to become immortal. On his death in February 2000, he was cryopreserved at Alcor, where he lies in cryogenic storage today.

We have thus seen that many transhumanists might advocate cryonics simply, so to speak, for practical purposes. While they may dream of a future unfettered by mortality, the reality of death is still of great concern in the present, and cryonics is the only available option of reaching the transhumanist future. While these considerations certainly play a role, I believe that the relationship between cryonics and transhumanism is not simply one of practical necessity. In my opinion, there are deeper causes relating to the fundamental outlook on life and personal purpose, which I will explore in the following paragraphs.

A shared feature of both movements is their rebellion against the natural limitations set by human frailty and death. As stated before, for transhumanists mortality is a grave bodily limitation that should be overcome, whereas for cryonics, the chance to cheat death is the core of all its efforts, as evidenced poignantly in the title of its founding manifesto, "The Prospect of Immortality". But the desire to overcome death by "unnatural" means does not suffice as a common explanation. Both the heart transplant and in vitro fertilization were reviled on similar grounds, without being regarded as specifically transhumanist topics. Indeed, cryonics may be seen as simply another (very speculative, very unconventional) medical procedure.

Examining the self-stated goals of both cryonics and transhumanism, we see several common points. For one, transhumanists generally have a strong belief in technical feasibility. The same is true of cryonicists: Most people simply would not pay large amounts of money for a procedure that is regarded as futile by the mainstream of the medical community without believing there to be a realistic chance for it to be successful.

Finally, we come to a point that is less obvious, but all the more significant: cryonics and transhumanism both take place in the far future. I believe that this observation is key to understanding their common attraction: they require a mental outlook that sees the far future as a place filled with desirable possibilities instead of threats and risks. For most people, the far future is not only uncertain, but it is also not of immediate concern: It takes a – very rare – kind of personal disposition to orient one's goals and actions not towards the problems at hand, but towards an uncertain and distant future beyond one's natural lifespan.

This brings us to our second question stated above: is the strong relationship between cryonics and transhumanism detrimental or beneficial for the cryonics movements?

It is easy to see how transhumanist ideas are detrimental: By expressing strong immortalist desires and grandiose visions for the future, early cryonicists alienated the scientific and medical communities, which might otherwise have helped establishing cryonics as a legitimate medical procedure. Instead, cryonics is currently a fringe movement and widely viewed as an eccentric and frightening Franken-science.

To understand if things might have turned out differently, let us imagine an alternative history in which cryonics developed independently of all transhumanist thought and ideals. In this hypothetical world, cryonics would only be a medical procedure, developed purely by scientists and doctors. Indeed, cryonics can be seen as the natural extension of the medical cooling techniques already established in emergency medicine and the nascent research into suspended animation, of which Blackstone et al. wrote in 2005:

"We report that hydrogen sulfide can induce a suspended animation-like state in a nonhibernating species, the house mouse (Mus musculus). This state is readily reversible and does not appear to harm the animal. This suggests the possibility of inducing suspended animation-like states for medical applications".

(As a side note, I will remark that contrasting the sober prose of this progress report with the grandiose visions quoted above makes clear why the cryonics community is not well-regarded in the medical establishment).

However, as appealing as our hypothetical history of a medical, professionalized cryonics might be to cryonics advocates, I conjecture that it simply could not have happened. The reason for this is simple: Cryonics is a medical procedure whose very viability will remain uncertain and inaccessible to empirical verification for decades at least. As such, it is hard to see how it could obtain any funding and support from academia.

Therefore I believe that without prominent transhumanist sponsors and the efforts of many transhumanists born from the sheer necessity to become active in cryonics for their own personal future, there simply would be no cryonics at all today.

This certainly does not mean things could not be better: The haphazard and improvised procedures prevalent in cryonics to this day have certainly not inspired confidence and cooperation from the medical community. Even if cryonics must remain outside of the recognition through the medical until the first patient is successfully reanimated, much could be done to foster a better relationship with scientists and doctors by adhering to established professional clinical standards in regards to documentation, procedures and hygiene.

§3. Conclusion

We have seen that not only is cryonics intimately linked with transhumanism, but that indeed transhumanist aspirations have been the

major force giving life to this speculative undertaking. Transhumanists need cryonics to reach the future, and thus contribute the
major portion of the effort, money and ideas needed to keep alive
the cryonics movement. This will change only at such a future time
when cryonics becomes just another medical procedure. For this
reason I believe that, for better or worse, the intimate relationship
between cryonics and transhumanism will endure. Perhaps not in
the distant, but certainly for the foreseeable future.

§4. Bibliography

Blackstone, E.A. et. al (2005). Hydrogen Sulfide Induces a Suspended Animation-Like State in Mice. Science. 308(5721):518.

Doug Bailey et. al (1998, updated 2002), Transhumanist Declaration. Available online at http://humanityplus.org/learn/transhumanist-declaration/

FM-2030 (1973), Up-Wingers: A Futurist Manifesto, John Day Co, ISBN
   0381900088

Hughes, James J. (2008), Report on the 2007 Interests and Beliefs Survey of
   the Members of the World Transhumanist Association. Available online
   at http://transhumanism.org/resources/WTASurvey2007.pdf

Kurzweil, Ray (2005), The Singularity is Near : When Humans Transcend
   Biology, The Viking Press, ISBN 0670033847

Roco, Mihail C. and Bainbridge, William Sims (eds., 2002), Converging
   Technologies for Improving Human Performance. Available online at
   www.wtec.org/ConvergingTechnologies/Report/NBIC_report.pdf

Cryonics in School Education

Jan Welke

For centuries the mindset that Aubrey De Grey termed "pro-aging trance" provided a plausible attitude, especially with a general life-expectancy that didn't exceed the late 50s.[1] The Industrial Revolution has led to significant improvement of the nutritional situation and that era saw major advancements in medical treatment and supply. And yet, these developments couldn't challenge the general conviction that aging and death would necessarily occur.

Today we have a fundamentally different situation as many of the formerly valid paradigms are being revised or called into question. Considering the chances that emerging life extension technologies and scientific progress in these fields offer to humanity it is rather astonishing that the potentials of these developments are all too often dismissed or even ignored. Cryonics is probably one of the best examples for an approach that aims to establish a connection to future life extension technologies and treatments. At present cryonics is probably the only technology that stops the process of decay taking place after a person dies. Unsurprisingly, this approach disconcerts many of the grown-ups and especially the elderly who have been brought up accepting death and aging as a natural necessity. However, it is quite astonishing to note that the reaction of many young people is quite similar.

When children are confronted with death, their reaction is usually characterized by resistance and refusal. Ordinarily, this happens for the first time when a loved pet dies or when they even lose a relative. Their first impulses are still very natural in so far that

---

[1] Aubrey DeGrey: Ending Aging. The Rejuvenation Breakthroughs That Could Reverse Human Aging In Our Lifetime. 2008. p. 16ff

they strongly wish to make this event undone by any means, even if no option to do so exists. Hence it is quite stunning to notice how exactly these children respond to this subject just ten years later. Most of them will see aging and death as way of the world that cannot be avoided from happening. The impression is created that they are echoing the convictions of preceding generations.

This raises the question when exactly the shift towards a death-accepting attitude takes place and which role public education at school plays in this context. There is evidence that the actual forming stage ends with puberty (age of 14-16). Aside from the parents, peer groups and school gain a growing influence on a young person's socialization and imprinting. The view of death as inevitable and natural is established in exactly these environments.

School education plays the key role here. But in the light of experience, as far as science and technology are concerned schools tend to be conservative to a high degree, in that what is taught in school is often lagging behind the current state of research. What used to be the latest state of research when teachers absolved their university studies has often seen adjustments due to new findings and discoveries. While some of them follow present developments in their fields with great attention, others stick to their outdated set of paradigms. Unfortunately, the latter is especially the case with teachers whose subjects are Biology, Chemistry and Physics. Their anti-cryonics and bio-conservative attitudes are in many regards on the same level as those of teachers of religious education (RE). In the latter case the anti-cryonics sentiment is usually to be found in strictly religious milieus and is rooted in the religion itself that claims a monopoly of interpretation in regard to death. Religious promises of an afterlife have the function to calm the fear of death of the believer and attempts to attain prolonged life in the here and now are oftentimes seen as hubris. This might not apply to Christian religion in general but is a wide-spread view. There is a similar situation with the eco-movement and other anti-progressive currents. The curriculum reflects these views in a moderated form in the same way that it suggests a general skepticism against research areas like genetic engineering.

RE is probably the only field in which a change towards a more open-minded view on cryonics is unthinkable. Science subjects however, usually reflect the general level of knowledge in a strongly simplified form and with considerable delay compared to the science-world. In so far, scientific progress in areas relevant to cryonics, or even in cryobiology or cryonics itself, will sooner or later entail adjustments in the respective curricula, rather later than sooner considering the strongly conservative nature of school and even more the committees that develop the curricula. And yet, it is exactly these committees that hold a key function in deciding how and if subjects like cryonics and life-extension are dealt with in school. Given that all the procedures involved in teaching are usually done in a way that shows a lot of subservience to authority (in this case the curricula-committees), the only way that change can occur is according to the top-down principle. Decision-making is the domain of those in the school board who are presiding the curricula-committees.

Still, it should be seen as progress that cryonics does already find brief mention here and there in school lessons, be it in English when dealing with teaching units such as science fiction or science and technology, or when discussing death and dying in ethic classes or in RE. With the exception of RE and science subjects, there is less of a categorical rejection, but still cryonics is usually treated as a sci-fi inspired oddity due to the teacher's personal bias. In consequence, the tendency is that school teaching contributes to intensifying prejudices against cryonics and generating skepticism.

Examined, the underlying assumptions that lead to opposing cryonics are for the most part identical with the misperceptions and misconceptions that characterize the majority view. First of all there is the common view shared by most people today who still regard aging and death as a natural process and inevitable process. In his work Ending Aging, Aubrey De Grey has identified and categorized a number of processes that constitute what is called aging and has pointed at possible ways to stop these from taking place or to reverse their effects.[2] The book has led to a clearer understanding of what actually makes the aging process and has shown that the damages in the single categories can be repaired by means of biotechnological engineering. It's not a question if the necessary treatments will be available in the future but rather when. In a far future, they may well be perfected to a degree that allows them to reverse the cause of death in a cryo-preserved person. It is also likely that computer technology will take on a major role in this context. Kurzweil points out the possibility of mind-uploading to be feasible once the singularity is achieved.[3] Yet, independent of whether and when this actually happens, the fact that cryonics is adjacent to advancements in various areas of science and technology leaves a broad scope of options that have the potential to reach cryonic's main objective: Restoring cryo-preserved patients back to life in good health and maybe in even better health than at any point of time in their youth.[4] Looking back in history there are good reasons to expect this to happen at some point in the future in a similar way in which physical damages that were lethal 50 or 100 years ago can be cured by routine treatments nowadays. Cryonics is oftentimes misinterpreted as a promising path immortality under false pretenses while in fact it is intended as a means to stop the processes of decay that occur in the human body after de-animation, hoping for future technology and medicine to offer the resources for successful reanimation. Whether or not cryonics will

---

[2] Aubrey DeGrey: Ending Aging. The Rejuvenation Breakthroughs That Could Reverse Human Aging In Our Lifetime. 2008.

[3] Ray Kurzweil: The Singularity Is Near – When Humans Transcend Biology. 2005. p. 5, 7-9

[4] Robert C. Ettinger: Man Into Superman. 2005.

succeed in defeating aging and death in the future will still have to be shown yet it is likely to contribute to this major aim by offering a bridge-technology. Thus cryonics is offering a chance, nothing more and nothing less. Once these misunderstandings are cleared up, a more of objective debate is possible.

In a pluralist society, the world view and the hopes of those who decide in favour of this chance (and against certain death) are a legitimate position and if furthermore technological progress in the related fields keeps accelerating as it has done so far, the visionary position may soon turned into a verified hypothesis. Attacking cryonics from a proclaimed moral position and a claim to absoluteness however is as unsuitable as it is a violation of basic principles of a democratic and pluralist society. With public schools obligated to the latter, one would at least expect neutrality here. As it is with other fundamental questions of life, the normative evaluation of cryonics and life extension should be up to the individual student. The diversity of opinions on this matter offers chances to practice discussion culture and mutual understanding.

Sources:

Aubrey De Grey: Ending Aging. The Rejuvenation Breakthroughs That Could Reverse Human Aging In Our Lifetime. 2008.

Robert C. Ettinger: Man Into Superman. 2005.

Ray Kurzweil: The Singularity Is Near – When Humans Transcend Biology. 2005.

II. Cryonics related topics

Organ differentiation and mortality

Klaus H. Sames

Summary

In multi cellular organs high specific performance is realized on the cost of regenerative processes. The study of structure and function in adult organs is able to reveal the limitations of regenerative capacities. Our hypothesis implies that, the inescapable mortality of highly developed multi cellular organisms has multiple organ specific causes. This means that the realization of the genetic program of organ development at the same time produces organ specific features, which limit the life span ("organ differentiation hypothesis").

Thus, the stop of aging or a rejuvenation is hampered by high complexity of mortality mechanisms. No break through is foreseeable for our next generations.

This situation leaves cryonics the only mean to profit from an extension of the life span, which hopefully will be possible in the future.

§ 1 Introduction

It is a common error to see different ways of life span extension as alternatives to cryonics.

1. Potential immortality may not be reached by the genera-
    tions living today without a cryonics storage. We must take
    into account the high complexity of facts leading to our
    mortality, whereby the latter becomes hard to influence.
2. A century of aging research has brought no break through
    (for aging s. chapter II).
3. There is also no emergency intervention in sight compara-
    ble to cryonics for people dying in an accident or by differ-
    ent pathologies today.

The origin of mortality must be of high complexity (compare e.g.
Jazwinski 2005). Our mortal state may be a direct consequence of
the development of organs showing complicated structures and
high performance. We must stop thinking about living entities like
a soup or cell culture or some arrangement of molecules, if we want
to understand our mortality; and we are to respect evolution.

The obligatory mortal state of highly developed multi cellular
organisms has multiple organ-specific causes, some of them well
known to each physician.

The aim of this paper is to demonstrate biological limitations of
the organ life span and to show the complexity of processes resulting in our mortality.

§ 2 The enigma of mortality

Human mortality and aging as caused on a complex organ level has
sparsely been investigated so far. Yet, our results and arguments
point just to this possibility. Life-maintaining mechanisms may be
interpreted as replacement, repair, or protection of organic molecules or supra molecular units. The main tool of such mechanisms
is the metabolism. Metabolism however, shows side effects. Therefore reduction of metabolism e.g. by cooling represents a protecting
effect acting against endogenous damage.

Under optimal conditions life maintaining mechanisms allow cell clones to grow indefinitely, as long as mitotic activity is guaranteed. During mitosis all components of the cell are reproduced at least in one of the daughter cells. Thus, there are no inevitable time-related changes in a number of undifferentiated systems, from cell clones up to low-developed multi cellular organisms as well as all germ lines. This is the warranty for the existence of life over millions of years. The germ lines may contain sexual recombination or may proceed without changes of DNA information by partheno-genesis (Collatz, 1994; Schartl et al. 1991; Schlupp et al. 1991; Smith-Sonneborn, 1987; Strehler, 1977). In other words, the avoidance of substantial molecular changes in a part of all living systems maintains the life on our planet. Cell division is the most effective anti aging repair mechanism.

Highly differentiated multi cellular animal systems however, show obligatory age-related changes and a species-specific limit of the lifespan. To explain this phenomenon Weisman postulated in 1892, that in such organisms, a potentially immortal germ line can be discerned from the somatic cells, which become mortal as a consequence of supra cellular organization. In high developed organisms aging starts just at the end of the growth period, when organs are optimally developed (a few exceptions with extended growth are under discussion). This seems to be paradoxical, because it is well known that multi cellular development leads to more effective life-maintaining mechanisms (e.g. immune reactions or blood circulation) by formation of organs. However, causes of mortality must have been established during development and natural selection of complex multi cellular organization and dependent cell differentiation phylogenetically as well as ontogenetically (compare Cutler et al.1978; Finch, 1993, Kirkwood, 1998).

Mortality may be a final consequence of increased specific performance, provided by complex organ structure, developed at the cost of regenerative and life-maintaining processes ("Organ differ-

entiation hypothesis": Sames, 1995; first formulation Sames 1980).
Similar suggestions can be found at R. Holliday (1995; 1998). This
implies that, mortality can not simply be explained on a molecular
level (Williams, 1957; compare Charlesworth, 1993).

As organ development had been driven by interspecies competition, repair of organs may have been neglected. This neglect allows a more rapid development of high-performance organs providing optimal reproduction, useful for coping with environmental
influences (including predators), and suited for fighting of competi-
tors. Thus, potential immortality may have been spent in favour of
rapid production of effective weapons for these struggles. The most
typical example is the rapid development of central nervous systems, the winners of this run. The number of post mitotic cells
seemingly has been steadily increased during brain evolution. More
and more maintenance and repair of the adult body has been replaced by the most powerful regeneration mechanism, reproduction. Consequently, the soma becomes disposable primarily in favour of high organ performance. This performance energy however, is not exclusively expended for reproduction in each species; it
may also be used for the defence of adult organisms. This can be
shown by birth control mechanisms in favour of food supply of
groups e.g. in wolf- and slender-tailed-mierkat families.

§ 3 Qualities of organs limiting the life span (examples)

The complexity of organ structure and function should imply the
mechanisms of mortality. Therefore studies of organs should be
suitable to reveal those mechanisms. In the following examples
organ analysis demonstrates weak points for possible endogenous
or exogenous damage attacks. The examples, which we have studied intensely in own experiments include rib cartilage, corneal endothelium and other eye tissues. We found, that these show features, which favor time-related changes. These observations can
possibly be generalized, since, using biomedical literature and

common medical knowledge, we later found examples for similar mechanisms (described below) in a number of other organs also.

Locomotory system

Age related changes of rib cartilage are at first found in its center, which also contains the oldest cells. Concentric matrix areas, which metabolically depend on a centrally located single cell or cell group represent the smallest living units of cartilage, named chondrons. Near to the cell, chondroitin sulfate forms the main side chains of matrix proteoglycans. It shows high turnover. Later on (starting during childhood) – seemingly at the point, when a critical volume of the chondron is reached, keratan sulfate forms at the outer zone of the chondron, far from the cell. Keratan sulfate shows lower turnover and a lower oxidation state as compared to chondroitin sulfate. Therefore it may be useful in locations at a critical distance from the cell, where metabolism is reduced by long diffusion ways for enzymes and substrates from/to the cell. In the cartilage center, the spaces of the chondrons reacting positively for chondroitin sulfate decrease in volume and the reaction becomes mixed with a positive reaction for keratan sulfate. Loss of proteoglycans can not be avoided and their concentration decreases in relation to age. Finally the total chondron undergoes so called "Verdämmerung" a fading-out without immediate degradation of the damaged or aged chondron. Two facts can explain this type of aging. a) the central cells are situated far from the perichondral blood vessels, while the cartilage itself is free of vessels. Vessels might be unable to bear pressure, while it is the main function of cartilage, to withstand high pressure load. The lack of vessels may cause hypoxia compensated partly by increased glycolysis. Thus, cell metabolism may become insufficient in many situations or decrease continually in chondrons as indicated by proteoglycan studies. b) new cartilage cells can only be produced inside the chondrons and cannot pass the matrix barrier to replace lost chondrons. Cells produced by the perichondrium are also unable to permeate the extra cellular matrix to replace lost chondrons. The metabolism of the increasing chondron free spaces in the center can only be guaranteed by neighboring chondrons, a fact possibly adding to metabolic overload of the remaining cells. On the other hand the diameter of rib cartilage seems to be maintained over the adult life span by restrictions of cell proliferation (Sames 1975; 1977; 1994a,b).

In aging bone, cells die in the outer parts of osteons (the smallest living units of bone) with increasing age. These parts are situated far from supporting vessels, which are found in the center of the osteon. Cell loss is favored by arteriosclerosis (Tonna,1977). Thus, similar to the situation in cartilage, substrate flow to the cells is diminished by the structure of the organ. However, it is not clear, why osteons are not replaced, while repair is possible in this tissue. One could speculate that bone regeneration might be connected with bone degradation and thus, the slow age-related loss of single cells, which does not affect overall bone stability, would not stimulate this first line of regeneration. Another explanation could be that cell loss exceeds the tissue replacement capacity.

Skeletal muscles are constructed by fibers which, as a consequence of their multinuclear structure, cannot divide by mitosis. They can be replaced by satellite cells, but these evidently do not replace aged fibers spontaneously. Possibly, these cells are not activated by the very slow age-related degeneration or loss of single fibers, while satellite cells remain active to some extend in aged muscles and can be activated by severe damage or hard training (Muradian and Frolkis, 2000). However, after stimulation of regeneration in aged muscles, insufficient up regulation of the Notch ligand Delta was found, which inhibited proliferation of satellite cells. Activation of Notch restored regenerative capacity (Conboy et al. 2003). One might ask, whether the prolonged resting of satellite cells could be a cause of such changes

Consequences for our organism: Aging of joint cartilage impairs functions of joints. Together with changes in bone and muscle besides those in the nervous system, immobility can result, which is crucial for life maintenance in old age. Aging of rib catilage contributes considerably to the decrease in thorax flexibility and thus, breathing capacity and oxygen supply to the brain.

This shows how time-related changes of one organ can cause mortality of a whole organism.

Neurosensory system

Corneal endothelial cells of the eye are taken to be mitotically inactive. This is proven by the fact that with increasing age, the flattened cells increase in diameter, while their density and number decrease. Cells from cattle up to 16 years of age showed the same or even higher proliferation velocity as compared to eyes from younger animals. The cells, at least in part, are therefore reversibly postmitotic cells (Sames 1989). The telomeres are well preserved in the aged cells and cells remain inter mitotic despite lack of telomerase (Egan 1998). Corneal endothelium represents a surface epithelium with border functions that maintains hydration status and thereby the transparency of the cornea by ion transport. It seems that denudation of the area covered by a cell is avoided by restriction of mitoses, since during mitosis the cell would be unable to cover its designated area of the basement membrane and this could lead to focal loss of transparency. Rather than by mitoses lost cells may be replaced more quickly by expansion of neighboring cells, since the cells show a high potency to expand laterally. Aqueous humor is the main source of cell support, and mitoses may be restricted by a lack of stimulating growth factors or the presence of inhibitory factors in this medium (Chen et al. 1999). Therefore cells cannot proliferate unless growth factors are supplied in cell culture or by bleeding injuries from the blood (Sames 1994a). Glycosaminoglycan monosaccharides, namely hexosamines and glucuronic acid, were used as parameters to reveal age-related changes and their dependence on mitotic activity. In primary cultures of aged eyes the amount of these substances was found to be reduced compared to young eyes. During culture an increase was observed and uronic acid reached the amount found in young eyes. However, young as well as aged cells showed in vitro aging (s. Sames and Lindner 1982).

The eye lens is growing appositionally like cartilage. Cell proliferation would lead to an increase of the organ volume. Therefore, mitotic activity must be reduced with age. The organ is supported by diffusion of nutrients from the aqueous humor. Similar to cartilage the oldest cells are situated in the depth of the tissue, where they may be insufficiently supported and stem cells seem to be unable to permeate and replace lens cells. Cells in the center loose their nuclei. Furthermore their proteins, lacking turnover, suffer radical damage, racemization, glycation, and crosslinking. In culture the area covered by each single lens cell increases, while cell density decreases with the number of passages, and overall oxygen consumption remains stable. Cells show increasing signs of degeneration and die after about 40 passages (Meyer, Sames 1981). Free radical production favored by light energy is involved in cross linking (Hockwin 1989; Kislinger et al. 2002). It is not known whether cells in the center loose the ability to perform mitoses because of light induced damage and why they loose cell nuclei. To replenish damaged cells and avoid an increase in size of the total lens, these cells would have to be emitted, but there seems to be a very limited cell recycling capacity in the aqueous humor of the eye.

Neurons can posses very long axons or high numbers of sophisticated dendrites and synaptic patterns inhibiting mitosis e.g. by inability to form a globular cell shape which is normally needed during mitosis. Neurons also contain very active mitochondria and part of the cells are overloaded by lipofuscin. Stem cells and astroglia cells are able to replace neurons forming functional neuronal networks (Berninger et al. 2007; Cameron and Gould, 1996; Hein-rich et al. 2011; Monje et al., 2003; Otto et al. 2009). Neuronal stem cells have also been described in the human brain (Fuchs et al., 2000; Sanal et al., 2004). Mezey et al (2003) found neurons with Y-chromosomes in brains of women after transplantation of male bone marrow. They interpret this finding as formation of neu-rons by bone marrow stem cells. The neurons observed had been small pyramidal cells in the cerebral cortex and granule cells in the hippocampus. Cells with processes of many centimeter lengths are hard to replace, since even during regeneration of large nerves the path to the target can be missed. Lastly however, embryonic stem cells have been induced to differentiate into pyramidal neurons forming connections with other neurons according to the layer into which they had been transplanted. Their axons ran in the correct tract to the periphery and some reached the pyramidal decussation and the descending spinal tract. Thus, correct replacement of corti-cal neurons by embryonic stem cells comes into sight (Ideguchi et al. 2010). The question remains again, why aging pyramidal neu-rons are not replaced spontaneously and develop increasing lipofuscinosis instead. Like other organs, the brain is unable to regenerate wholly at an advanced age and healing often involves glial scar formation. Furthermore, the question is still open, wheth-er stem cells can maintain the memory function of the cells, they replace. During stress – a physiological performance of high devel-oped organ systems – hippocampal neurons can become damaged (e.g. Sapolsky et al., 1986; Sawada et al., 2004).

We do not need explain the severe consequences of neuron loss and cataract for the aging organism.

Circulatory System

In blood vessels the constant blood pressure load leads to frequent pathological changes and possibly to increased cell turnover of arterial endothelial cells as indicated by arteriosclerosis. Shortening of telomeres may result (Chang and Harley 1995). This is also indicated by the comparison of veins (low dynamic blood pressure) and arteries (high blood pressure) (s. at Sames 1993). Treatment of rats of different ages with angiotensin converting enzyme inhibitor to reduce blood pressure had the effect of delaying the thickening of intima and media of arterial walls (Michel et al., 1994). In human legs upright walking causes high hydrostatic pressure in veins, which may lead to varicose veins (s. at. Sames 1993). On the other hand changes of the atherosclerotic type are not found in veins

Heart muscle cells form a network which shows continuous motion (contraction, stretching) in which mitoses may endanger stability. As in most organs stem cells have been found in the heart muscle in niches with reduced motion. Transdifferentiation into functioning cardiomyocytes of these cells is in discussion. (Sussman and Aversa, 2004; Timmermanns et al. 2003). Contractile tissue with a number of properties of muscle tissue can be generated from embryonic stem cells. External stem cells injected into heart muscle lesions are able to form contractile cardiomyocytes. It is unclear if motion is an obstacle for integration of stem cells localized in the heart and external stem cells (Befahr et al. 2002; Jackson et al. 2001; Malouf et al. 2001; Orlic et al. 2001; Zimmermann et al. 2006). It remains to be shown that stem cells can be integrated into intact aged heart muscle tissue to replace or support aged cardiomyocytes. Postmitotic cardiomyocytes are at the same time overloaded by a high number of mitochondria with high oxidative activity. Under such structural and functional conditions free radicals (which seem to present no problem e.g. for cells of the germ line) may lead to damage, as suggested by increasing contents of lipofuscin in heart muscle. Furthermore blood pressure may favor fibrosis in the heart as shown by comparison of the valves of the right and left heart (Keller et al. 1999; Keller et al. 2001). One also finds asymmetrically localized metabolic age-related changes. E. g. in the left ventricle of aged rats the angiotensin receptor ATIa-

and ATIb-mRNA levels are upregulated in comparison to young rats, while in the right ventricle no such change can be found (Heymes et al., 1998). A decrease in procollagen type I mRNA has been observed in the left but not in the right ventricle of rat hearts (Annoni et al. 1998).

The consequences of changes in the circulatory system and their contribution to mortality are well known.

Urogenital System

Kidney aging shows that complicated structures like renal glomeruli cannot be replaced. Yet, age-related loss of a high number of glomeruli (e.g. Bürger 1960) may be partly avoidable, as it was found to be minimal in centenarians (Studies of W. Selberg, Hamburg, oral communication).

In ovaries elastoid fibrosis of blood vessels occurs during each ovulation (ovulation sclerosis) (Loewe, 1973). An antagonistic role in ovulation has been postulated for Interleukin-1beta, which starts the initial inflammation phase and TGF-beta initiating fibrosis and healing as the second step (Derman et al. 1999).

Especially the loss of glomeruli may lead to organ failure of kidneys. The changes in the ovary contribute to menopause and it's hormonal disturbances

Respiratory system

The lung suffers irreversible accumulation of small particles from the air leading to time-dependent increase of damage. Furthermore, particles are stored in the lymphatic system including lymph nodes outside of the lung at the hilum and may invade blood vessels.

This situation is not compatible with an indefinite function of this organ of essential vital importance. The changes may contribute to emphysema or atelectases in high age and thus, to part of age-related mortality on the population level.

Digestive System

In the liver stem cells maintain mitotic capacity but do not seem to undergo mitoses during a normal (even an age-related) functional state. Liver cells show signs of mitotic restriction namely hyperploidia and lipofuscin accumulation and the pool of intermitotic cells decreases with increasing age (Muradian and Frolkis, 2000; Zeeh, 1991). One can speculate that mitotic activity of normal liver cells is restricted, due to lack of a complete cell recycling system including cell degradation, since an increase of cell number cannot be permitted in an encapsulated organ of limited size. This is in agreement with the fact, that mitotic activity is high (at least in stem cells), following damage or injury with extended loss of cells. The non dividing cells may loose mitotic capacity due to causes, that are not fully understood up to now. It is not clear if single cells are lost with increasing age and if they can be replaced by stem cells.

Impairment of liver function by age-related changes may be no cause of death but contribute to dysregulations of detoxification and to difficulties in excretion of substrates.

Whole body: limitation of the body size

In general, following the completion of the growth period mitotic activity is only allowed for cell replacement. Surface epithelia have no problem to exclude aged cells. Sophisticated cell or tissue recycling – similar to the one found in blood cell systems or bones – would be needed in most other tissues to allow for mitotic activity without further organ growth.

In the adult human organism mitotic activity is primarily found in stem cells and the precursor cells of different tissues. According to Kohn (1975) 90% of mammalian cells show mitotic inactivity after the stop of growth. The question is allowed if a reduction of cells – able to perform cell divisions – to such an extend could allow at all the regularly replacement of all body cells. There exist no postmitotic cells or organisms with an unlimited (non-aging) lifespan. Cell loss is difficult to compensate for without mitosis. On the other hand mitosis is the most effective regeneration mechanism of the cell.

Interestingly life span extending interventions may be connected to reduced numbers of cell divisions (stem cells and other inter mitotic cells s. de Magalháes, Faragher 2008).

§ 4 Conclusion

The organ examples show limitations of regenerative mechanisms, which are the immediate consequences of organ structure and function themselves. In part they lead to cell loss, which may be reparable by stem cells.

There may however, be mechanisms, which cannot be influenced by stem cells and would need total destruction and reconstruction of the tissue, which is partly realized in bones, even if bones in the adult human organism seem to be never rejuvenated as organs in toto.

The time related changes of mortal organs caused by their immanent restriction of repair may be the basis of aging mechanisms. Different secondary influences modulate the mortality mechanisms. The result is a more complex, time related change called aging.

Functions and structures of tissues, organs and organ systems as well as their onto- and phylogenesis are an integral part of mortality. The examples treated above contain phenomena which explain mortality of our organism. However, the enumeration is not a complete list of events that cause mortality, nor does it represent the secondary reactions. Mortality cannot be reversed without changes in essential structures and functions of our organs, unless we learn how to replace damaged elements cell by cell and molecule by molecule of a total organism. All intervention strategies available today provide only limited modification of the lifespan in short-living animals. Application to human beings has not yet been tested. Their influences on basal aging processes remain speculative since the nature of these processes is hypothetical. One should be aware that, some extension of the lifespan does not mean an influence on primary aging mechanisms in each case (Bernarducci, 1996).

As a consequence of the complexity of mortality processes and their dependence on organ structure and functions we have no conventional means like pharmacological treatment, avoidance of damage, enhancement of production of essential substances lacking in high age etc., which could stop aging totally or rejuvenate our organism.

§ 5 Bibliography

Annoni, G., Luvara, G., Arosio, B., Gagliano, N., Fiordaliso, F., Santambrogio, D., Jeremic, G., Mircoli, L., Latini, R., Vergani, C., Masson, S. 1998. Age-dependent expression of fibrosis-related genes and collagen deposition in the rat myocardium. Mech. Ageing Dev. 101, 57-72.

Befahr, A., Zingmann, L.V., Hodgson, D.M., Rauzier, J.M., Kane, G.C., Terzic, A., Puceat, M. 2002. Stem cell differentiation requires a paracrine pathway in the heart. FASEB J. 16, 1558-1566.

Bernarducci, M. P., Owens, M.J. 1996. Is there a fountain of youth? A review of current life extension strategies. Pharmacotherapy 16, 193-200.

Berninger, B., Costa, M.R., Koch, U., Schroeder, T., Sutor, B., Grothe, B., Götz, M. 2007. Functional properties of neurons derived from in vitro reprogrammed postnatal astroglia. J. Neurosci. 27, 8654-8664; doi:10.1523/JNEUROSCI.1615-07.2007.

Bürger, M. 1960. Altern und Krankheit als Problem der Biomorphose. Thieme, Leipzig.

Cameron, H.A., Gold, E. 1996. Distinct populations of cells in the adult dentate gyrus undergo mitosis or apoptosis in response to adrenalectomy. J. Comp. Neurol. 369, 56-63.

Charlesworth, B. 1993. Evolutionary mechanisms of senescence. Genetica 91, 11-19.

Chang, E., Harley, C. B. 1995. Telomere length and replicative aging in human vascular tissue. Proc. Natl. Acad. Sci. USA 92, 11190-11194.

Chen, K.H., Harris, D.L., Joyce, N.C. 1999. TGF-beta 2 in aqueous humor suppresses S-phase entry in cultured corneal endothelial cells. Invest. Ophthalmol. Vis. Sci. 40, 2513-2519.

Collatz, K.G. 1994. Unbegrenzt lebensfähige Systeme. In: Olbrich, E., Sames, K., Schramm, A. (eds.) Kompendium der Gerontologie, IV-1. Ecomed Verlag, Landsberg, 1-16.

Conboy, I.M., Conboy, M.J., Smythe, G.M., Rando, T.A. 2003. Notch-mediated restoration of regenerative potential to aged muscle. Science 302, 1575-1577.

Cutler, R.G. 1978. Evolutionary biology of senescence. In: Behnke, J.A., Finch, C.E., Moment, G. E. (eds.): The biology of aging. Plenum Press, New York, 311-360.

Danner, D.B., 1992. The proliferation theory of rejuvenation. Mech. Ageing Developm. 65, 85-107.

Derman, S.G., Kol, S., Ben-Shlomo, I., Resnick, C.E., Rohan, R.M., Adashi, E.Y. 1999. Transforming growth factor-beta 1 is a potent inhibitor of interleukin-1beta action in whole ovarian dispersates. J. Endocrinol. 160, 415-423.

Egan, C.A., Savre-Train, I., Shay, J.W., Wilson, S. E., Bourne, W.M. 1998. Analysis of telomere length in human corneal endothelial cells from donors of different ages. Invest. Ophthalmol. Vis. Sci. 39, 648-653.

Fuchs, E., Kempermann, G., Winkler, J., Kuhn, H. 2000. In: Sames, K. (ed.): Medizinische Regeneration und Tissue Engineering. Ecomed Verlag, Landsberg, V-9, 1-8.

Heinrich, C., Gascón, S., Masserdotti, G., Lepier, A., Sanchez, R., Simon-Ebert, T., Schroeder, T., Goetz, M., Berninger, B. 2011. Generation of subtype-specific neurons from postnatal astroglia of the mouse cerebral cortex. Nature Protocols 6, 214–228.

Heymes, C., Silvestre, J.S., Llorens-Cortes, C., Chevalier, B., Marotte, F., Levy, B.I., Swynghedauw, B., Samuel, J.L. 1998. Cardiac senescence is associated with enhanced expression of angiotensin II receptor subtypes. Endocrinology 139, 2579-2587.

Hockwin, O. 1989. Physiologisches Altern demonstriert am Beispiel der Augenlinse. In: Baltes, M., Kohli, M., Sames, K. (eds.) Erfolgreiches Altern Bedingungen und Variationen. Verlag Hans Huber, Bern, 240-247.

Holliday, R. 1995. Understanding ageing. Cambridge: Cambridge Univ. Press.

Holliday, R. 1998. Causes of aging. Ann. New York Acad. Sci. 854, 61-71.

Ideguchi, M., Palmer, T.D., Recht, L.D., Weimann, J.M. 2010. Murine embryonic stem cell-derived pyramidal neurons integrate into the cerebral cortex and appropriately project axons to subcortical targets. J. Neuroscience 30, 894-904.

Jackson, K.A., Majka, S.M., Wang, H., Pocius, J., Harley, C.J., Mjesky, M.W., Entman, M.H., Michael, L.H., Hirschi, K.K., Goodtell, M.A. 2001. Regeneration of ischemic cardiac muscle and vascular endothelium by adult stem cells. J. Clin. Invest. 107, 1395-1402.

Jazwinski, S.M. 2005. Genetics of longevity from yeast to human. In: Sames, K., Sethe, S., Stolzing, A.: Extending the life span. Lit Verlag, Münster, Berlin, London, pp 133-139.

Keller, F., Pflug, A., Lehmann, Y. 2001. Veränderungen der kollagenen Strukturen menschlicher Herzklappen in Abhängigkeit von Alter und Hypertonie. Z. Gerontol. Geriat. 34, 470-475.

Keller, F., Werner, R., Wähner, J., Köhler, T., Wolff, W., Leutert, G. 1999. Zur histologischen Biomorphose menschlicher Herzklappen II. Morphometrische Untersuchungen. Z. Gerontol. Geriat. 32, 104-111.

Kirkwood, T.B.L. 1998. Biological theories of aging: An overview. Aging Clin. Exper. Res. 10, 144-146.

Kislinger, T., Schneider, M., Pischetsrieder, M. 2002. Accumulation of non-enzymatic glycation products on proteins and DNA. In: Oehmichen, M., Ritz-Timme, S., Meissner, C. (eds.) Aging. Morphological, biochemical, molecular and social aspects. Schmidt-Römhild, Lübeck, 235-248.

Kohn, R.R. 1975 Intrinsic aging of postmitotic cells. In Blandau, R.J. (ed.) Aging Gametes. New York, Karger, 1-18.

Loewe, K.R. 1973. Untersuchungen über die Ovulationssklerose am Gefäßsystem menschlicher Ovarien. Habilitationsschrift, Mannheim Fakultät für Klinische Medizin Mannheim der Universität Heidelberg.

Malouf, N.N., Coleman, W.B., Grisham, J.W., Lininger, R.A., Madden, V.J., Sproul, M., Anderson, P.A. 2001. Adult-derived stem cells from the liver become myocytes in the heart in vivo. Am. J. Pathol. 158, 1929-1935.

Mayer, U.M., Sames, K. 1981. Sauerstoffverbrauch und Alterung in Langzeitkulturen von Linsenepithel. Albrecht v. Graefes Arch. Klin. Ophthalmol. 217, 117-124.

Mezey, E., Key, S., Vogelsang, G., Szalayova, I., Lange, G.D., Crain, B. 2003. Proc. Natl. Acad. Sci. 100, 1364-1369.

Michel, J..B., Heudes, D., Michel, O., Poitevin, P., Philippe, M., Scalbert, E., Corman, B., Levy, B. I. 1994. Effect of chronic ANG I-converting enzyme inhibition on aging processes. II. Large arteries. Am. J. Physiol. 267, R124-R135.

Monje, M. L., Toda, H., Palmer, T.D. 2003. Inflammatory blockade restores adult hippocampal neurogenesis. Science. 302, 1760-1764.

Muradian, K., Frolkis, V. 2000. Regeneration und Altern. In: Sames, K. (ed.) Medizinische Regeneration und Tissue Engineering. Ecomed Verlag, Landsberg, V-3.1-V-3.20.

Orlic, D., Kajstura, J., Chimenti, S., Jakoniuk, I., Anderson, S.M., Li, B., Pickel, J., McKay, R., Nadal-Ginard, B., Bodine, D.M., Leri, A., Anversa, P. 2001. Bone marrow cells regenerate infarcted myocardium. Nature 410, 701-705.

Otto, F., Illes, S., Opatz, J., Laryea, M., Theiss, S., Hartung, H.P., Schnitzler, A., Siebler, M., Dihné, M. 2009. Cerebrospinal fluid of brain trauma patients inhibits in vitro neuronal network function via NMDA receptors. Ann. Neurol. 66, 546-555.

Sames, K. 1975. Histochemical studies on the distribution of acidic glycosaminoglycans in human rib cartilage during the aging process. Mech. Ageing Dev. 4, 431-448.

Sames, K. 1977. Einfluß des Knorpeldurchmessers auf die Entwicklung degenerativer Veränderungen im Zentrum alternden menschlichen Rippenknorpels. Act. Gerontol. 7, 555-561.

Sames, K. 1980. Altern durch Fehlkonstruktion? Umschau 80, 181-182.

Sames, K. 1989. Proliferation behaviour of endothelial cells from bovine cornea aged in vivo. In: Niedermüller, H. (ed.) Advances in Experimental Gerontology. Facultas Universitätsverlag, Vienna, 93-103.

Sames, K. 1993 Zur Physiosklerose der Vene. Phlebologie 22, 218-221.

Sames, K. 1994a. The role of proteoglycans and glycosaminoglycans in aging. Basel: Karger.

Sames, K. 1994b. Glycosaminoglycans, proteoglycans and aging. In: Jollès, P. (ed.) Proteoglycans. Birkhäuser, Basel, 243-274.

Sames, K. 1995. Naturwissenschaftliche Kausalgerontologie In: Olbrich, E., Sames, K., Schramm, A. (eds.) Kompendium der Gerontologie III/5, Ecomed Verlag, Landsberg, 1-48.

Sames, K., Lindner, J. 1982. Changes in cell cultures of bovine corneal endothelium cells as related to donor age and number of passages in vitro. Akt. Gerontol. 12, 206-212

Sanal, N., Tramontin, A.D., Quinones-Hinojosa, A., Barbaro, N.M., Gupta, N., Kunwar, S., Lawton, M.T., McDermott, M.W., Parsa, A.T., Verdugo, M.G., Berger, M.S., Alvarez-Buylla, A. 2004. Unique astrocyte ribbon in adult human brain contains neural stem cells but lacks chain migration. Nature 427, 740-744.

Sapolsky, R.M., Krey, L.C., Mc Ewen, B.S. 1986. The neuroendocrinology of stress and aging: the glucocorticoid cascade hypothesis. Endocr. Rev. 7, 284-301.

Sawada, T., Morinobu, S., Tsuji, S., Kawano, K., Watanabe, T., Suenaga, T., Takahashi, T., Yamawaki, S., Nishida A. 2004. A Reduction in levels of amphiphysin 1 mRNA in the hippocampus of aged rats subjected to repeated variable stress. Neuroscience. 126, 461-466.

Schartl, M., Schlupp, I., Schartl, A., Meyer, M.K., Nanda, I., Schmid, M., Epplen, J.T., Parzefall, J. 1991. On the stability of dispensable constituents of the eukaryotic genome: stability of coding sequences versus truly hypervariable sequences in a clonal vertebrate, the amazon Molly, Poecilia formosa. Proc. Natl. Acad. Sci. 88, 8759-8763.

Schlupp, I., Parzefall, J., Schartl, M. 1991. Male mate choice in mixed bisexual/unisexual breeding complexes of Poecilia. Ethology 88, 215-222.

Smith-Sonneborn, J. 1987. Longevity in the protozoa. Basic Life Sci. 42, 101-109.

Strehler, B.L. 1977. Time cells and aging. Academic Press, New York.

Sussman, M.A., Anversa, P. 2004. Myocardial aging and senescence: where have the stem cells gone? Ann. Rev. Physiol. 66, 29-48.

Timmermanns, F., De-Sutter, J., Gillebert, T.C. 2003. Stem cells for the heart, are we there yet? Cardiology 100, 176-185.

Tonna, E.A. 1977. Aging of the skeletal-dental systems and supporting tissue. In: Finch, C.E., Hayflick, L. (eds.) Handbook of the Biology of Aging. Van Nostrand Reinhold Comp, New York, 470-495.

Weisman, A. 1892. Über Leben und Tod. In: Aufsätze über Vererbung und verwandte biologische Fragen. Fischer, Jena.

Williams, G.C. 1957. Pleiotropy, natural selection, and the evolution of senescence. Evolution 11, 398-411.

Zeeh, J. 1991. Leber. In: Platt, D. (ed.) Biologie des Alterns. Walter de Gruyter, Berlin, 246-54.

Zimmermann, W.H., Didie, M., Doker, S., Melnychenko, I., Naito, H., Rogge, C., Tiburcy, M., Eschenhagen, T. 2006. Heart muscle engineering: An update on cardiac muscle replacement therapy. Cardiovasc. Res. 71, 419-29.

General mechanisms of mortality and aging and their relation to cryonics

Klaus H. Sames

Summary

Negative consequences of complex body structure and function lead to the mortal state of an organism. They are modulated by exogenous and endogenous damage like free radicals, high temperature, toxic substances, metabolic errors or expression of adverse genes etc., which limit the life span. On the other hand the latter may be determined by genetic processes in the adult organism, which modulate those primary time-dependent adverse changes, but are unable to reverse them. The result of these interwoven processes represents aging. Thus aging is of much higher complexity as compared with mortality processes. This implies that a stop of aging or rejuvenation cannot be reached by conventional means like drugs or gene therapy. We need methods in development now like rejuvenation of adult stem cells and their transplantation as well as nano technology. These technologies will need much time for their development. The study of aging corroborates a fact already uncovered by observations in mortality mechanisms namely that, the only technology to allow for preservation of life – not perfect itself – today is cryonics.

Introduction

The organ examples treated in the first chapter ("Organ differentiation and mortality") provide causes of restriction of tissue repair as well as the replacement of damaged parts in favor of organ structure and function.

In general, lack of replacement and repair must lead to losses and damage increasing with time, whereby irreplaceable elements can be lost with and without aging (Gavrilov and Gavrilova 2003).

We shall treat causes of mortality as a general precondition for the impact of the stochastic as well as system immanent damaging events resulting in processes called "aging".

Since mortality cannot be totally reversed by all means available today, as stated before, our only chance remains cryonics. To explain this fact, we should further study, on a general level, the primary causes of mortality and the superimposed mechanisms of aging.

Mortality and aging not only make cryonics indispensable, they also influence its mechanisms and procedures.

Living organisms are compromises between demands of construction and the desirable functions. Such compromises may tend to labilities counterbalanced by different regulation mechanisms in a living organism. If regulations themselves are impaired, a total brake down may result in general organ failure with increasing decay ("death"). Total structural reconstruction hopefully will restore the functional equilibria (revival).

Lack of cell division and the consequences

Severe age-related changes, which have been taken to initiate the aging process, mainly occur in post mitotic cells or growth arrested (confluent) culture cells. Causes inhibiting mitoses are explained in part I. Damaged macromolecules and organelles (biological garbage, waste) increase when even autophagic degradation (by lysosomes, proteasomes and cytosolic proteases) is inhibited. In dividing cells however, mortality is reduced in comparison to growth arrested cells even if one suppresses autophagic degradation in both cell types (Stroikin et al 2005).

Mutations increase in relation to age in tissues containing postmitotic cells (Ono et al., 2002). Damage in mitochondria is also strictly organ-specific and shows changes, which are most pronounced in postmitotic cells, mitotically inactive cells or cells with high oxygen turnover (Brunk and Terman, 2002; Meissner et al., 2001; Miquel, 1998, Wei et al., 2001). In contrast Rasmussen et al. (2003) have found no substantial decrease in functional parameters of human skeletal muscle mitochondria.

Glycation of DNA has also been shown in post mitotic cells (Schneider et al. 2004). Some types of post mitotic cells are exposed to function-specific metabolic overload or mechanical overstress such as high numbers of mitochondria with high oxygen turnover in the cell. As discussed above, increasing lipofuscin content which is mediated by free radicals may be a consequence, while in inter-mitotic cell populations lipofuscinosis is marginal or not present at all and cell loss is reversible. An example are the testes, where inter-mitotic germ cell clones are free of lipofuscin, while postmitotic Leydig cells show pronounced lipofuscinosis (Miquel, 1998; Rune et al., 1991).

Restrictions of mitoses or cell replacement may be effected e.g. by shortened telomeres or inhibited proliferation mechanisms as well as lack of growth factors.

It may even be that, lack of mitoses itself leads to changes, which reduce the ability of the cell to perform division (Danner 1992) or other functions. An important factor of rejuvenation by mitosis may be asymmetric division as shown in yeast, where one cell retains the accumulated damag, while the other is young, seemingly containing de novo produced material (Jazwinski 2005).

The example of the eye lens (s. above) reveals lacking replacement of cells, as a consequence of changes in molecular structures by insufficient metabolism, and at the same time damage by exposure to light energy. Macromolecules in non dividing cells are susceptible to age-related changes (Ritz-Timme, 2002), as postulated by Verzar (1965) and Bürger (1960). Replacement of cells may be impossible by a matrix or other structure which does not allow for cell migration (cartilage, eye lens).

Restriction of mitotic activity inhibits also another one of the most powerful life maintaining mechanisms: the selection of damaged cells. Thus, apoptosis cannot lead to an increase of vitality in tissues with restricted mitosis. Only where stem cells are available, apoptosis may be beneficial as part of a system replacing damaged cells by stem cells. Even without aging, postmitotic cells can be lost in accidents, disease, unspecific damage (e.g. temporally increased local temperature or metabolite concentration), or by the fact that, global regulation systems of the organism do not register the fate of single cells.

Loss of cells exceeding replacement capacities

The number of cells may decrease e.g. by excessive cell loss or lack of replacement. The viability of cells may decrease if they are not recycled in due time intervals. In inter mitotic cells, cell loss may exceed mitotic capacity. Examples may be endothelia of blood vessels (s. part I) or other cell layers lining surfaces. In post mitotic cells the capacity of replacement by stem cells may be exceeded by cell loss.

Increased cell loss by regulations disrespecting the demand of cells

It is commonly known that, nutrition, mitotic activity, and metabolic activity of cells in a multi cellular organism depend on 'public' demands and may be regulated against vital interests of single cells. In a multi cellular differentiated system, cells depend on other cells. For example the pH of the extra cellular matrix may be influenced by blood composition and the local cell may be unable to regulate it in favor of the enzyme optima needed for the turnover of matrix constituents. As far as we know, general regulation systems of the organism are able to control such parameters as blood pH, but are unable to control single cell parameters. It seems that general regulations of our organism neither respect the health of a single cell nor register its loss even if it is impossible to replace the cells concerned. This causes no harm where cells can be readily replaced.

Consequences of the homeothermic body temperature

Homeothermic temperature is beneficial to certain organ functions, but it also leads to destruction of macromolecules. This damage can be repaired but may increase vulnerability and initiate irreversible changes (Setlow, 1987). Especially post mitotic cells may suffer slowly increasing damage.

Influences of low temperature are treated in the last part: "extension of the life span and cryonics".

Increased cell loss by restriction of supply for cells and tissues as a consequence of organ construction

The organ examples show that there exist restrictions:

of supply by reduced flow of metabolites in a specialized matrix (cartilage, vessel wall?),

of supply for the cells in areas far from supporting blood vessels (cartilage, bone),

of nutritional supply for the cells by a medium poor in nutrients (eye),

of blood supply in favor of function (s. chapter I, transparency of eye, pressure load in cartilage)

General difficulties of blood supply exist at the borders of neighboring capillary beds, a fact, well known to pathologists. This may lead to cell loss, while cells may be hard to replace in such locations. One could speculate that in regions with low blood flow or higher distance from blood vessels, physiological metabolites accumulate and reach toxic concentrations. Similarly, in areas with high metabolic activity reduced blood flow may lead to increasing temperature.

Increased cell performance by high volumes of extra cellular matrix

A high mass of extra cellular matrix substances can increase metabolic load of cells which metabolize these substances and may, at the same time, be hard to control by the cells (s. cartilage), since diffusion ways for all substrates and enzymes are very long, making metabolic processes more time consuming.

The role of stem cells

Stem cells provide the main source of replacement of post mitotic cells. While stem cells are able to produce new muscle fibers, liver cells etc. they seem unable to rejuvenate aged organs by spontaneous processes.

The question remains open, if cells in complicated tissue interconnections are replaceable this way and how stem cells can be differentiated and integrated correctly. Facts mentioned in the organ examples show inhibitions of stem cell migration or stem cell mediated repair in some tissues (s. e.g. the lung and kidney). Therefore even without aging of stem cells the human organism remains a mortal one. Earlier studies indicated that, purified undifferentiated hematopoietic stem cells show an unlimited proliferation potential in vitro (Barnes 1988; Spangrude et al. 1988). It remains to clear if stem cells are involved into aging by primary aging of structures in their environment (niche), or – in contrast to the earlier findings – by endogenous changes. Many markers of cellular "aging" have been described in stem cells e.g. markers of proliferation, signaling, and telomere length. Thus, stem cell "aging" is discussed as a causal factor for loss of parenchymal cells (Conde, Streit 2006; Geiger et al. 2007; Sethe et al, 2005; Stolzing et al. 2008). In adult stem cells telomere shortening and dysfunction can be found with induction of DNA checkpoints leading to cell cycle arrest and/or cell death (Ju, Rudolph 2008). P53 is involved into selection of aging stem cells with telomere dysfunction. Its deletion increases the life span of chromosomal-instable intestinal stem cells (Begus-Nahrmann et al. 2009; Sharpless, De Pinho 2007). Stem cell aging may also be a secondary phenomenon of aging. Since stem cells are direct descendents of unlimitedly growing germ cell lines, the causes of their aging have to be explained. It remains an open question if rejuvenated adult stem cells or even embryonic stem cells could repair age-related changes all over the organism.

Lack of program governed regeneration

Adult tissue structures of high complexity in composition or form obviously cannot be replaced if lost or damaged. Renal glomeruli cannot be replaced possibly due to problems with their vascularization/drainage and – may be – the lack of a construction program that coordinates and joins tissue elements of different evolutionary origin (a program like that occurring during embryonic development).

In principle, adult organisms may be able to repair or replace blood vessels, but such abilities are not used, or cannot be used for full restoration or replacement e.g. of an aged or sclerotic aorta with changes of its three dimensional form. Causes may include difficulties with elimination of the aged vessel, lack of space for the production of a new vessel, reconstitution of continuity of circulation without leakage, and lack of a program governing the building of large three-dimensional structures in its original shape in an adult organism.

Insufficient tissue regeneration: scarring and fibrosis

There exist physiological processes (ovary), pathologies or accumulations of external waste (lung s. part I.) leading to fibrosis. Scarring as well as fibrosis resulting from incomplete regeneration mechanisms, represent irreversible changes (studies of complete regeneration mechanism of e.g. total limbs are ongoing in amphibia s. e.g. Tsonis et al. 1995). Thus, mechanical stress, injuries, inflammations, tissue damage by different causes, and age-related specific diseases in mammals all may lead to the same final product: fibrosis, which increases with increasing age (Sames, 2004). To remove it, degradation of the matrix and replacement of cells would be needed. In some cases rearrangement of the 3 dimensional order may also be a problem.

Restricted catabolic metabolism

In molecules bearing mechanical load (e.g. fiber proteins) the catabolic side of turnover (degradation) may be reduced to guarantee mechanical stability of organs. In the case of collagen fibers slow metabolism leads to age-related fibrosis and at the same time to cross linking e.g. by glycation (s. at Sames 2004). Collagen amount increases as it can be produced, but not degraded to the same extend. Again it seems that, cellular metabolic criteria are not the only ones determining the turnover of proteins and their posttranslational changes over a prolonged time.

Evolution

Antagonistic pleiotropy theories of aging predict that, functions and structures as well as the genes concerned have positive effects during the reproductive phase of life. However they may become deleterious during the post reproductive phase. Reproduction is more important for a species than the maintenance of an individuals post reproductive life (Kirkwood, 1998; Kirkwood, 2002; Kirkwood, Austad 2000). During evolution cells or organisms become disposable with the first development of replication mechanisms. The general notion is that, organs of high complexity may be easier reproduced than repaired, since more and more sophisticated repair mechanisms would be needed with increasing complexity of organs during development.

The organ differentiation hypothesis represents an antagonistic pleiotropy theory of mortality, as high performance organs with insufficient repair are favorable in younger age but undergo increasing changes over the long run. The hypothesis explains also the intra species and organ dependent variation in aging, since instabilities causing mortality at the same time open the door for all sorts of stochastic influences. Another somewhat sophisticated explanation of intra species variability, by Martin (2009) proposes genetically governed randomization in gene expression, allowing for adaptation to the environment, which persisting leads to pathological changes with increasing age, an antagonistic pleiotropic process.

Aging

At first we are to discern mortality and aging. Mortality means that we will die inescapably some time or the other even without a disease or accident. Aging represents a pattern of time related events, including those caused by mortality mechanisms, somehow comparably evolving in each human being. The colloquial term "aging" is defined by the time related changes visible in members of the human species. Beyond this no clear cut generally accepted definition of "aging" exists. There exists also no evidence for the identity of human "aging" with time related changes in other species. One of the most important markers of aging, the life span is an argument against interspecies identity. Can changes needing 120 years to kill a mammal be identical with those killing another mammal after 3 years? In the frame of this article the substantial interspecies differences of aging cannot be treated in detail.

Mortality causes an increasing vulnerability of the organism. Its processes seem to be modified by secondary, ubiquitous, endogenous and exogenous, detrimental influences, leading to regular time related changes. These are modified by tertiary stochastic influences as well as adaptations and regulations of the organism mitigating the consequences. There may be e.g. gene expressions acting against damaging influences or stabilizing organisms. The resulting complex process is called aging. In turn aging contributes to mortality (for age-related changes s. Finch and Hayflick 1977; Spence, 1989). By manipulation of external damaging influences, the life span may be increased to a limited extent.

Interpretation of total human aging on a molecular level would therefore be a similar task to interpreting a football match on the molecular level. It is an open question if life extension methods available today – allowing for limited life span extension – really mean an intervention into the basal aging process or the primary causes of mortality. Such interventions are of limited effect since aging is not stopped by any method available today.

External and secondary causes contributing to "aging"

Many time related changes and influences on the life span have been studied in very different animals assuming that aging is the same process in all of them. This however cannot be proven since there exists no simple definition of aging up to now.

A number of molecular theories like the free radical theory of aging describe secondary events and thus, do not explain the origin of aging. For example in cartilage lack of oxygen, retarded flow of metabolites and the absence of cell replacement caused by the structure of cartilage are more likely explanations of time related changes in cartilage than free radicals in this hypoxic tissue. Cells are comparable to stoves composed of combustible materials. Their anti oxidation mechanisms are crucial for their existence and only pathological disturbances of these mechanisms allow for the damaging influence of free radicals (comp. De Magalhães, Church 2006).

Shortening of telomeres may also be a secondary event contributing to aging. In the adult organism germ line cells and stem cells show levels of telomerase, which would be high enough to maintain the length and function of telomeres without temporary limitation (von Zglinicki T. et al. 2005). Most of our other cells are mitotically inactive ones which suffer no shortening of telomeres by cell division. Telomere shortening (von Figura et al. 2009) seems to be a secondary or tertiary event in mortality as it can be a consequence of oxidative stress or disease (Houben et al. 2008; Lansdorp 2009; Terry-Adkins et al. 2006).

High energy expenditure, toxic side-effects of hormones, endotoxic and exotoxic agents, protein cross linking, changes in signal systems, changes in irreplaceable proteins (Pichetsrieder 2003), metabolic errors, and apoptosis are to be mentioned in this context.

Additionally a host of genetic factors has been considered, such as genes with adverse effects, insufficient genetic repair, changes in methylation or – permanently in discussion – accumulation of mutations (which interestingly, mainly occurs in post mitotic cells and is organ specific). In fact an increasing number of genes are known to be related to the lifespan. Gene activities and mutations prevalently studied in short lived animals are connected to an increase or decrease of the life span by different mechanisms (de Magalhães, Faragher 2008).

Basal metabolic rate, body size, brain size, (calendar) age at maturity, and postnatal growth rate all show some correlation to longevity (de Magalhães et al. 2007). Classical signaling pathways and transcription factors related to longevity have been studied e.g. TOR, AMP kinase, sirtuins and IGF-1 (Kenyon 2010). For instance a high number of genes are involved into the control of oxidative or immunological reactions etc.. Genetic mechanisms may not govern aging processes or causes of mortality, but improve the utilization of resources during the curse of aging or at least in younger years. The same holds true with cellular theories e.g. in vitro aging or programmed aging. Nonetheless, they may describe important modulators of the causal aging changes in an already mortal organism. Theories propagating changes of single organs as primary causes of aging also give no explanation of the real causes of aging in the respective organ itself. E.g. the immunological theory of aging corroborated by T cell aging, relies on complicated interactions between the constituents of the immune system (comp. Pawalec et al. 2010) allowing no clear cut analysis of the primary causes of immunosenescence (for reviews and details of aging theories see: Adelman and Roth 1982; Burzynski 2003; Harman 2003; Harman et al. 1998; Jazwinski 1996; Klebanov et al. 2001; Lane

2003; Mikhelson 2001; Ono et al. 2002; Patridge 2001; Sames 1995; Warner et al., 1987; Yeo 2002).

The limited proliferation potential of inter mitotic cell clones may be a secondary adaptation in organisms with limited lifespan. It may be caused by the surroundings of cells in an aging organism, while in cell cultures it may be the consequence of highly artificial conditions.

Transformation of cells, cloning research and studies of reproduction and regeneration show that, differentiated cells can be rejuvenated to an extent, which is not yet fully known. However, we may assume that many age-related changes are reversible by deprogramming and reprogramming and therefore are not fixed on the cellular level (Gurdon et al., 1979; Hu et al., 1977; Monk, 2002). E.g. cell fusion in part of the studies has rejuvenated the resulting cells (s. at Stolzing et al. 2007).

The different aging theories and facts have been treated earlier in detail (Sames 1995) and cannot be repeated in the frame of this chapter. Molecular age related changes are organ specific rather than generalized. This can be demonstrated for example by the results of proteoglycan research (Sames, 1990; Sames, 1994a and 1994b) and points to organ specificity of aging mechanisms. Other examples for organ specificity of age related molecular changes can be found in mitochondria especially of post mitotic cells or during replicative senescence of culture cells (Meissner et al. 2005; Passos, Zglinicki 2005) as well as in age differences of anti oxidative defense mechanisms (Martin et al. 2005).

Genetic mechanisms of irreversible time related changes

The organ differentiation hypothesis explains genetic influences as follows: Activity of genes involved in the differentiation program (Finch, 1993) of complex multi cellular systems lead – by the formation of our organs – to a general limitation of the lifespan and to the mortal condition of the organism developed. Other genes not discussed in detail in this paper, may be related to longevity via expressions e.g. serving performances of metabolism, stress resistance, chromatin-dependent gene regulation, and genome stability (Jazwinski 2005) and may modify the lifespan in a species-specific way. In short-lived animals, there may be genes, which favor senescence and become active in high age (which is not usually reached under normal environmental conditions). Mutations in such genes can increase the lifespan under optimum conditions (s. e.g. Finch, 1993). Since expressions of the cell genome are differentiated organ-specifically, generalizations can result in simplification where organ-specific aspects are neglected.

Extension of the life span and cryonics

We are to realize that testing of life span extension needs at least 70 years in human beings. Starting with 50 (where hopefully age related changes can still be influenced) one has to wait until an age higher than 123 years has been reached. If success is missing, you are to start another longtime experiment. We are also to remember that, life span extension may not mean influencing the basic aging processes.

Today attention must be given to regenerative medicine including cell and tissue replacement, gene technological reconstitution of essential elements and nano technology. Organ transplantation could be used as a mean of rejuvenation, but aging starts again in the transplant and we would need repeated severe operations. It would also imply transplantation of nearly each organ of the body. A number of such operations would bear high risk.

Replacement of neurons by exact integration of stem cells into the neuronal network reveal the potency of stem cell methods. However, it is a complicated process including signaling an aged cell to enter apoptosis, signaling neighboring cells or macrophages to ingest the remains of the apoptotic cell, signaling stem cells to migrate to the place of the apoptotic cell, signaling the stem cell to differentiate into a cell of the same type as the original cell. To rejuvenate the total organism by stem cell manipulation or - transplantation is far from realization. It has to be proven if transplanted external stem cells are able to repair aged tissue. Rejuvenation by thousands of cell transplantations seems not to be imaginable.

The removal of aged cells and replacement by stem cells is not possible in each organ. Some organs could only be regenerated by total degradation and replacement of all structures e.g. cartilage, eye lens or bone. In heart muscle destruction of aged muscle cells may only be possible after integration of stem cells to bridge the gap in the network of connected muscle cells, which is in steady motion.

It would seem to be overdoing e.g. to deconstruct/reconstruct the total lung to get rid of deposed external particles, because macrophages are unable to transport it to the outside, but in fact we know no better way.

In cases, where increased load or stress (e.g. blood pressure) lead to damage, repair processes would need to reach a speed equal to that of damage.

If irreplaceable elements are lost (e.g. glomeruli) even such type of repair must fail. One is e.g. unable to imagine a way of renal regeneration without reinstallation of the embryonic structure and the embryonic development program or to use nano technology.

The chance to replace all aging elements continually maintaining an organism in a juvenile state by artificial procedures, seems also to be far away. The retrospect indicates that one may have to think in many decades. The worth of cryonics consists in its potential ability to bridge this time.

As a consequence of such complications we may have to wait for the advent of a sophisticated nano technology able to replace damaged tissues molecule by molecule working at some billions of sites in the human organism in parallel, guided by an external three dimensional molecular image of the human organism.

However, all types of interventions by repair mechanisms should lead to immediate reaction. The results could be easily controlled without stating a life span extension with its high consumption of time.

Aging and cryonics

Cryobiology and cryonics will play a crucial role in life extension. On the other hand knowledge of age related changes influences methods of cryonics e.g. perfusion:

- of sclerotic vessels,
- of aged lungs with high resistance of pulmonal vessels or
- of parts of circulation with increased resistance by loss of capillary vessels or alveoli etc.

Cryonics is a field of thermo biology at deep subzero temperatures. A general principle of temperature depression is reduction of metabolism, leading to increasing independence of cells of substrate supply and excretion, down to temperatures, where we observe no changes over thousands of years.

In poikilotherms low environmental temperature can increase the life span by reduction of metabolism (Lamb 1977). The body temperature of mice can be lowered by diet with a concurrent increase of their life span (Weindruch, Walford 1988).

Hibernators show higher life spans as compared to comparable non sleepers (Reitz 1996). It has been widely accepted that the human species is unable to hibernate caused by its tropical origin. However insect eaters near to the roots of primates like hedgehogs and even tropical tenrecs are hibernators. Furthermore tropical primates also are able to hibernate. This was shown with the Madagascan dwarf lemurs e.g. the fat-tailed dwarf lemur Cheirogaleus medius (Dausmann et al. 2004, 2009; s. at Fiedler 1970).

However, the problem to depress the temperature of hibernators to cryogenic temperatures without harm to the animal is far from being solved.

Similar to mortality cryo preservation depends on organ structure, which is not only responsible for mortality, but also influences cryopreservation.

We should inter al. reflect the fact that, our tissues are full of natural cryoprotectants. For example the matrix of cartilage contains Glycosaminoglycan molecules binding a volume of water 300 times higher compared to their own one. Cartilage matrix is very resistant to freezing (own observation not published). The glycosaminoglycans are effective and nontoxic cryoprotectants. The possibility to increase endogenous cryoprotectant levels e.g. by inhibitors of excretion to my knowledge has not been tested so far.

The most serious problem of body structure is the high mass of the human body. This should play no role during cooling by perfusion with cool solution, since the overwhelming part of cells are reached in nanometer distance from the capillaries. However here we face the same problems, which lead to cell aging in some structures, namely the high distances of a number of cell groups from capillaries in a number of organs.

Furthermore the capillary bed provides short cuts allowing wide areas to be circumvented by perfusion. Hopefully the brain circulation knows no functionally dead zones of perfusion as a consequence of the high demand of brain cells in oxygen and glucose. In heart surgery a perfusion with a 10°C solution or blood via the carotids causes no harm.

Blood vessels may be hard to conserve by cryopreservation if not isolated from their tissue surroundings (comp. Hamilton et al. 1973).

Some cell types show special age related changes, especially post mitotic cells. Such cells may need more perfect careful perfusion, a factor already respected in brain cells by other causes. Many irreplaceable cells can be identified by their content of lipofuscin.

Aging and age-related loss of brain cells endanger the outcome of cryonics as doe postmortem ischemia and freezing damage. Thus, the mortality problem of restricted replacement of cells effects cryonics directly.

The goal is saving as much of life bearing residues as cryonics can manage to do. The long time conception commonly agreed to, is a breakthrough in cryonics allowing to cryopreserve human bodies before aging and dying.

Cryobiology will be needed for regenerative medicine and thus, for suspension repair. Besides cryo preservation of stem cells for repair it is also able to develop storage of engineered tissue and transplant organs over the long run.

Cryonic resuspension makes no sense without rejuvenation and repair of patients organisms. We may hope that, repair of one organ following the other will be possible by stepwise thawing. This means that organs should be stable following the thaw. Otherwise a great team or sophisticated techniques must allow repair of different organs before or after the thaw by organ specific different methods simultaneously. An alternative may be the development of very rapid thawing methods.

Today most important of all life maintaining methods undoubtedly is the most perfect cryopreservation available, leaving other problems to future development.

Conclusion

Aging research is needed as well as cryonics to reach a substantial extension of the life span, where cryonics provides the only mean of life extension at hand today.

Bibliography

Adelman, R. C., Roth, G. S. (ed.) 1982. Testing the theories of Aging. CRC Press Inc., Boca Raton.

Barnes, D.M. 1988. Blood-forming stem cells purified. Science 241, 24-25.

Begus-Nahrmann, Y., Lechl, A., Obenauf, A.C., Nalapareddy, K., Peit, E., Hoffmann, E., Schlaudraff, F., Liss, B., Schirmacher, P., Kestler, H., Danenberg, E., Barker, N., Clevers, H., Speicher, M.R., Rudolph, K.L. 2009. P53 deletion impairs clearance of chromosomal-instable stem cells in aging telomere-dysfunctional mice. Nature Genetics 41, 1138-1143.

Brunk, U.T., Terman, A. 2002. The mitochondrial-lysosomal axis theory of aging: accumulation of damaged mitochondria as a result of imperfect autophagocytosis. Eur. J. Biochem. 269, 1996-2002.

Buerger, M. 1960. Altern und Krankheit als Problem der Biomorphose. Thieme, Leipzig.

Burzynski, S.R. 2003. Gene silencing – a new theory of aging. Med. Hypotheses 60, 578-583.

Conde, J.R., Streit, W.J. 2006 Microglia in the aging brain. J. Neuropathol. Exp. Neurol. 65, 199-203.

Danner, D.B. 1992. The proliferation theory of rejuvenation. Mech. Ageing Developm. 65, 85-107.

Dausmann, K.H., Glos, J., Ganzhorn, J.U., Heldmaier, G. 2004. Physiology: hibernation in a tropical primate. Nature 429, 825-826.

Dausmann, K.H., Glos, J., Heldmaier, G. 2009. Energetics of tropical hibernation. Journal of Comparative Physiology B 179, 345-357.

De Magalhâes, J.P., Church, G.M. 2006. Cells discover fire: employing oxygen species in development and consequences for aging. Exp. Gerontol. 41, 1-10.

De Magalhâes, J.P., Costa, J., Church, G.M. 2007. An analysis of the relationship between metabolism, developmental schedules, and longevity using phylogenetic independent contrasts. #J Gerontol A Biol Sci Med Sci. 62, 149-160.

De Magalhâes, J.P., Faragher, R.G. 2008. Cell divisions and mammalian aging: integrative biology insights from genes that regulate longevity. Bioessays 30, 567-578.

Fiedler, W. 1979. Die Herrentiere. In: Grizmeks Tierleben, Vol. 10 Kindler publisher, Zuerich, p 259.

Finch, C.E. 1993. Theories of aging. Aging Clin. Exp. Res. 5, 277-289.

Finch, C.E., Hayflick, L. (eds.) 1977. Handbook of the biology of aging. Van Nostrand Reinhold Comp, New York.

Gavrilov, L.A., Gavrilova, N.S. 2003. The quest for a general theory of aging and longevity. Sci. Aging Knowl. Environ. 28, p. re5.

Geiger, H., Koehler, A., Gunzer, M. 2007. Stem cells, aging, niche, adhesion and Cdc42. A model for changes in cell-cell interactions and hematopoietic stem cell aging. Cell Cycle 6, 884-887.

Gurdon, J.B., Laskey, R.A., De Robertis, E.M., Partington, G.A. 1979. Reprogramming of transplanted nuclei in amphibia. Int. Rev. Cytol. Suppl. 9, 161-178.

Hamilton, R., Holst, H.I., Lehr, H.B. 1973. Successful Preservation of Canine Small Intestine by Freezing. Journal of Surgical Research 14, 313–318.

Harman, D. 2003. The free radical theory of aging. Antioxid. Redox Signal 5, 557-561.

Harman, D., Holliday, R., Mohsen, M. (eds.) 1998. Towards Prolongation of the Healthy Life Span. Ann. New York Acad. Sci. 854, 61-67.

Houben, J.M.J., Moonen, H.J.J., van Schooten, F.J., Hageman, G.J. 2008. Telomere length assessment: biomarker of chronic oxidative stress? Free Radic. Biol. Med. 44, 235-246.

Hu, F., Pasztor, L.M., Teramura, D.J. 1977. Somatic cell hybrids derived from terminally differentiated rhesus cells and established mouse cell lines. Mech. Ageing Dev. 6, 305-318.

Jazwinski, S. M. 1996. Longevity, genes and aging. Science 273, 54-59.

Jazwinski, S.M. 2005. Genetics of longevity from yeast to human. In: Sames K., Sethe S., Stolzing A.: Extending the life span. Lit Verlag, Münster, Berlin, London, pp 133-139.

Ju, Z., Rudolph, K.L. 2008. Telomere dysfunction and stem cell ageing. Biochimie 90, 24-32.

Kenyon, C.J. 2010. The genetics of aging. Nature 464, 504-512.

Kirkwood, T.B.L. 1998. Biological theories of aging: An overview. Aging Clin. Exper. Res. 10, 144-146.

Kirkwood, T.B. 2002. Evolution of ageing. Mech. Ageing Dev. 123, 737-745.

Kirkwood, T.B.L., Austad, S.N. 2000. Why do we age? Nature 408, 233-238.

Klebanov, S., Astle, C.M., Roderick, T.H., Flurkey, K., Archer, J.R., Chen, J., Harrison, D.E. 2001. Maximum lifespans in mice are extended by wild strain alleles. Exp. Biol. Med. 226, 854-859.

Lamb, M.J. 1977. Bioloy of ageing. Blackie, London, pp. 76-79.

Lane, N. 2003. A unifying view of ageing and disease: the double agent theory. J. Theor. Biol. 225, 531-540.

Lansdorp, P.M. 2009. Telomeres and disease. EMBO J. 28, 2532-2540.

Martin, G.M. 2009. Epigenetic gambling and epigenetic drift as an antagonistic pleiotropic mechanism of aging. Aging Cell 8, 761-764.

Martin, R., Fitzl, G., Mozet, C., Welt, K., Wieland, E. 2005 Can antioxidants influence the life span? In: Sames K., Sethe S., Stolzing A.: Extending the life span. Lit Verlag, Münster, Berlin, London, pp 107-115.

Meissner, C., Mohamed, S.A., von Wumb, N., Oehmichen, M. 2001. Das mitochondriale Genom und Altern. Z. Gerontol. Geriatr. 34, 447-451.

Meissner, C., Storm, T., Buse, P., Oehmichen, M. 2005. Fragmentation of mitochondrial DNA and the aging process. In: Sames K., Sethe S., Stolzing A.: Extending the life span. Lit Verlag, Münster, Berlin, London, pp 141- 153.

Mikhelson, V.M. 2001. Replicative mosaicism might explain the seeming contradictions in the telomere theory of aging. Mech. Ageing Dev. 122, 1361-1365.

Miquel, J. 1998. An update on the oxygen stress-mitochondrial mutation theory of aging: genetic and evolutionary implications. Exp. Gerontol. 33, 113-126.

Monk, M. 2002. Mammalian embryonic development –insights from studies on the X chromosome. Cytogenet. Genome Res. 99, 200-209.

Ono, T., Uehara, Y., Saito, Y., Ikehata, H. 2002. Mutation theory of aging, assessed in transgenic mice and knockout mice. Mech. Ageing Dev. 123, 1534-1552.

Passos, J.F., von Zglinicki, T. 2005. Mitochondria, telomeres and cell senescence. Exper. Gerontol. 40, 466-472.

Patridge, L. 2001. Evolutionary theories of ageing applied to long lived organisms. Exper. Gerontol. 36, 641-650.

Pawelec, G., Larbi, A., Derhovanessian, E. 2010. Senescence of the human immune system. J. Comp. Pathol. 142 Suppl.1, 39-44.

Pischetsrieder, M. 2003. Prevention of advanced glycation end products (AG-ES) in proteins. Presentation given at the conference "Extending the lif span." Hamburg 24.-26.September 2003.

Rasmussen, U.F., Krustrup, P., Kjaer, M., Rasmussen, H.N. 2003. Experimental evidence against the mitochondrial theory of aging. A study of isolated human skeletal muscle mitochondria. Exp. Gerontol. 38, 877-886.

Reitz A: In Alters Frische, Verlag Gesundheit publisher, Berlin 1996, p 267.

Ritz-Timme, S. 2002. Protein modifications and aging. In: Oehmichen, M., Ritz-Timme, S., Meissner, C. (eds.) Aging. Morphological, biochemical, molecular and social aspects. Schmidt-Römhild, Lübeck, 261-274.

Rune, G.M., De-Souza, P., Merker, H. J. 1991. Ultrastructural and histochemical characterization of marmoset (Callithrix jacchus) Leydig cells during postnatal development. Anat. Embryol. 183, 179-191.

Sames, K. 1980. Altern durch Fehlkonstruktion? Umschau 80, 181-182.

Sames, K. 1990. Age-related changes of morphological parameters in hyaline cartilage. In: Robert, L., Hofecker, G. (eds.), The Theoretical Basis of Aging Research. Facultas Universitätsverlag, Vienna, 177-184.

Sames, K. 1994a. The role of proteoglycans and glycosaminoglycans in aging. Basel: Karger.

Sames, K. 1994b. Glycosaminoglycans, proteoglycans and aging. In: Jollès, P. (ed.) Proteoglycans. Birkhäuser, Basel, 243-274.

Sames, K. 1995. Naturwissenschaftliche Kausalgerontologie. In: Olbrich, E., Sames, K., Schramm, A. (eds.) Kompendium der Gerontologie III/5, Ecomed Verlag, Landsberg, 1-48.

Sames, K. 2004. Altern, Fibrose und Reaktionsmechanismen des Bindegewebes. In: Ganten, D., Ruckpaul, K., Ruiz-Torres, A. (eds.) Molekularmedizinische Grundlagen von altersspezifischen Erkrankungen. Springer Verlag, Heidelberg, 402-428.

Schneider, M., Thoß, G., Hübner-Parajsz, C., Kientsch-Engel, R., Stahl, P., Pischetsrieder, M. 2004. Determination of glycated nucleobases in human urine by a new monoclonal antibody specific for N2-carboxyethyl-2'-deoxyguanosine. Chem. Res. Toxikol. 17, 1385-1390.

Sethe, S., Scutt, A., Stolzing, A. 2006. Aging of mesenchymal stem cells. Ageing Res. Rev. 5, 91–116.

Setlow, R.B. 1987. Theory presentation and background summary. In Warner, H.R., Butler, R.N., Sprott, R.L., Schneider, E.L. (eds.) Modern biological theories of aging. Aging Vol. 31, Raven press, New York, 177-182.

Sharpless, N.E., De Pinho, R.A. 2007. How stem cells age and why this makes us grow old. Nature Reviews Mol. Cell Biol. 8, 703-713.

Spangrude, G.J., Heinfeld, S., Weisman, I.L. 1988. Purification and characterization of mouse hemopoietic stem cells. Science 241, 58-62.

Spence, A. P. 1989. Biology of human Aging. Prentice-Hall Inc., Englewood Cliffs New Jersey.

Stolzing, A., Hescheler, J., Sethe, S. 2007. Fusion and regenerative therapies: is immortality really recessive? Rejuvenation Res. 10, 571-586.

Stolzing, A., Jones, E., McGonagle, D., Scutt, A. 2008 Age-related changes in human bone marrow-derived mesenchymal stem cells: consequences for cell therapies. Mech. Ageing Dev. 129, 163-173.

Stroikin, Y., Dalen, H., Brunk, U.T., Terman, A. 2005 Testing the "garbage" accumulation theory of ageing: mitotic activity protects cells from death induced by inhibition of autophagy. Biogerontology 6, 39-47.

Tarry-Atkins, J.L., Ozanne, S.E., Norden, A., Cherif, H., Hales, C.N. 2006. Lower antioxidant capacity and elevated P53 and p21 may be a link between gender in renal telomere shortening, albuninuria and longevity. Am. J. Physiol. Renal Fluid Electrolyte Physiol. 290, F509-F516.

Tsonis, P.A., Washabaugh, C.H., Del Rio-Tsonis, K. 1995. Transdifferentiation as a basis for amphibian limb regeneration. Sem. Cell Biol. 6, 127-135.

Verzar, F. 1965. Experimentelle Gerontologie. Ferdinand Enke, Stuttgart.

Von Figura, G., Hartmann, D., Song, Z., Rudolph, K.L. 2009. Role of telo-mere dysfunction in aging and its detection by biomarkers. J. Mol. Med. 87, 1165-1171.

Von Zglinicki, T., Martin-Ruiz, C.M., Saretzki, G. 2005. Telomeres, cell senescence and human ageing. Signal Transduct. 3, 103-114.

Warner, H.R., Butler, R.N., Sprott, R.L., Schneider, E. L. (eds.) 1987. Modern biological theories of aging, Vol. 31. New York, Raven Press.

Wei, Y.H., Ma, Y.S., Lee, H.C., Lee, C.F., Lu, C.Y. 2001. Mitochondrial theory of aging matures – roles of mtDNA mutation and oxidative stress in human aging. Zhonghua Yi. Xue. Za. Zhi. (Taipei) 64, 259-270.

Weindruch, R., Walford, R.L. 1988.The retardation of aging and disease by dietary restriction. Charles C Thomas Publisher, Springfield p 260.

Yeo, E.J., Park, S.C. 2002. Age-dependent agonist-specific dysregulation of membrane-mediated signal transduction: emergence of the gate theory of aging. Mech. Ageing Dev. 123, 1563-1578.

The Ice Age in Cardiac Surgery and Rescue Medicine

Holger Zorn

Summary

Cold is, for centuries, applied medicine: to treat fever, to reduce blood loss, as a mean of anesthesia, and to delay cell death. Deep hypothermia made intra cardiac procedures under circulatory arrest possible. Extra corporeal circulation with moderate hypothermia is standard in cardiac surgery. Mild hypothermia improves survival after out-of-hospital cardiac arrest and becomes state-of-the-art therapy in rescue medicine.

§ 1. Ancient times: from 4th Century BC to 19th Century

The use of hypothermia was already described by the ancient Egyptians, Chinese, Greeks and Romans. Hippocrates advised to pack severely wounded patients into ice and snow to reduce blood loss.[1] Robert Boyle, the Irish natural scientist, suggest an ice-cold brine bath to treat patients with typhoid fever.[2] However, the development of induced controlled clinical hypothermia is closely connected with the development of cardiac surgery. Therefore, a look back helps to understand its todays role.

In 1883, Theodor Billroth, one the world's most famous surgeons, has been told about the presentation of a living dog with a sutured cardiac wound at the 11th International Medical Congress in Rome, and answered: "A surgeon who tries to suture a heart wound deserves to lose the esteem of his colleagues."[3] However, a few years later could be stated by Dr. Sherman: "The direct route to the heart is only two or three inches long, but the cardiac surgery took 2,400 years to traverse it ".[3] What had happened? The first successful operation of a human heart was done by Ludwig Rehn in Frankfurt, Germany, on September 9, 1896. The ice was broken, a rapid development began and led to the ice age in cardiac surgery.[4]

Two French scientists have first systematically clinically explored and described the underlying principles: Jean-Dominique Larrey (1766–1842), military physician, personal physician to Napoleon Bonaparte and inventor of the "flying ambulances"; and his contemporary Julien Jean Cesar Legallois (1770–1814), physiologist and lecturer at the Paris Medical School. Larrey noted in a "Memorandum on the dry gangrene caused by cold" during the Poland campaign, that Freshly operated soldiers survived better if they had to sleep away from the fire in the snow. In addition, those patients had as less pain the colder it was.[5] Legallois' experimental work with decapitated and perfused rabbits led him to the conclusion: If one could supply a heart continuously with arterialized blood, no matter if natural or artificial, there's no doubt that live may be extended endless.[6]

§ 2. Two competing concepts: hypothermia and perfusion

Based on these fundamental works, two surgeons, Bill Bigelow and John Gibbon, have developed two competing concepts for the upcoming surgical discipline cardiac surgery. William Gordon Bigelow (1913–2005) worked at Toronto University Hospital and Branson Research Institute. Encouraged by Larrey's notice, he focused his research on hypothermia primarily as a means of anesthesia and examined the effect of cold on the metabolism of groundhogs. His technique consists of an cathode-ray electrocardiograph, an electric stimulator, a thermometer, an oxygen tank, a cooling blanket that covers the animal, a refrigerating machine, a blood pressure manometer, and a diathermy re-warming cabinet.[7,8] As a side effect, he has developed the first pacemaker, to overcome the ventricular fibrillation that occurs regularly during rewarming. All in all, a very fascinating life's work, written by himself and published 1984 as "Cold Hearts".[9] However, Bigelow did not have the privilege be the first to try out his technique in humans. This was done by Lewis and Taufic at the University of Minnesota in September 1952, September 2nd. They closed an atrial septal defect of a 5 years young female patient. The body was cooled for 90 minutes down to 26°C / 79 F, followed by 58 minutes operation time – including 6 minutes cardiac arrest – and 35 minutes rewarming time.[10]

John Heysham Gibbon (1903–1973) worked at Jefferson Medical College, Philadelphia. Assisted by his wife and granted by IBM, he developed the principles of the artificial circulation, inspired by Legallois' prediction. Model number three, the first well-working heart-lung machine was heavy, huge and had a very sophisticated inner life with pumps and tubes and valves and connectors, but the principle was quiet simple and is still the same today: An artificial circuit is designed to remove blood from the patient's circulation, deliver that blood to a blood oxygenation and purification device, and then return the arterialized blood to the patient while the patient's heart and lung are clamped. John Gibbon did the first successful open heart procedure in human on May 6th 1953: He closed a large atrial septal defect in an 18-year-old college girl.[11]

§ 3. The Drew technique – Combined concepts

At the end of that decennium, Charles Drew from Westminster Hospital, London, published a unique perfusion technique: Taking into account that heart and lungs are connected in series, he used quadruple cannulation and two coupled pumps to bypass only the heart, to keeps the lung in, thus to avoid an artificial oxygenator. A very sophisticated technique that has found only limited distribution. With the means of the time it was quiet impossible to synchronize the left heart pump and the right heart pump in a way, to avoid a volume overload of the atrial cavities. But Drew and his team brilliantly handled this technique. Their first patient was a 12 month old mongoloid infant with multiple abnormities. The preparation for bypass has taken about 45 minutes, and then they started the left heart bypass at 150 ml/min and cooled the blood. 15 minutes later, the pharynx temperature dropped from 36°C down to 26°C, the heart had slowed considerably and the right heart bypass was started too. The pump flow was set at 450 – 500 ml/min to keep a mean blood pressure of 80 mmHg, and they continued cooling down in 27 minutes. When the temperature was 15°C they stopped both pumps and performed the real operation with an arrest time of 45 minutes.[12]

Consequently, combining both concepts (hypothermia and perfusion) was only a matter of time. An industrially mass-manufactured disposable, Bentley's Temptrol oxygenator, appeared on the market which contained a bubble oxygenator and a stainless steel heat-exchanger in one housing.[13]

§ 4. Modern times: current tools in hypothermic heart surgery

Actually, a modern two-tank heater-cooler device may cool water in three minutes from 36 °C down to 18 °C in about three minutes. And a state-of-the-art oxygenator, e. g. PrimOx from Sorin Group in Mirandola, Italy, with only 0.14 m² temperature exchange surface is able to use this cooling power and can cool down the 5–6 liters blood down to 18°C in approximately 20 minutes, depending on his own performance factor, calculated by good approximation as

$$\eta = (Tbout - Tbin) / (Twin - Tbin)$$

with Tbout – blood outflow temp, Tbin – blood inflow temp, Twin – water inflow temp, and Tbin – blood inflow temp. Today, this is clinical routine with roughly one million procedures per year worldwide. However, evolved surgical techniques have led to the practice of mild instead of profound hypothermia, thus cooling down to 32–34 °C, in most cardiac centers.

§ 5. Mild therapeutic hypothermia in rescue medicine

This technique, mild hypothermia, has now become an independent therapeutic tool to cool patients who have survived a cardiac arrest due to ventricular fibrillation. Background was the frustrating fact that 9 out of 10 people suffering sudden cardiac arrest will die within six month, although 5 out of 10 could be successfully primarily resuscitated. Gerald Buckberg, the inventor of cold blood cardioplegia, has once commented that by saying "We save the heart but we loose the brain". However, since 2003 there is evidence that induced, controlled mild hypothermia increases the survival rate and quality of successfully resuscitated patients after cardiac arrest: Michael Holtzer from Vienna University Hospital, and the Hypothermia After Cardiac Arrest Study Group (HACA) published the results of 137 cooled versus 138 conventionally treated patients. Seventy-five (55%) of the patients in the hypothermia group had a favorable neurologic outcome, as compared with 54 (39%) in the normothermia group. Mortality at six months was 41% in the hypothermia group, as compared with 55% in the normothermia group and with non-significant differences in complication rates between the groups.[14] Since 2003, mild hypothermia has been included in various guidelines, starting with the International Liaison Committee on Resuscitation (ILCOR) and recently clarified by the European Resuscitation Council (ERC).[15,16]

Driven by ILCOR, ERC and national councils, the technique of mild therapeutic hypothermia has been widely accepted and adapted – and has become more invasive. Our own program, started at Martin Luther University Hospital in Halle (Saale), Germany, in 2006, used an invasive cooling catheter, inserted into the femoral vein and advanced into the vena cava inferior. A roller pump (Stockert Instruments GmbH, Munich, Germany) with pressure and temperature module was combined with a heater-cooler device and mechanically modified in a way that a rotation speed of 25 min-1 cannot be exceeded. A sterile tubing set with a single use heat exchanger (Vision, EuroSets Srl, Medolla, Italy) was connected with an endovascular-cooling catheter (ICY Intravascular Catheter, Alsius Corp., Irvine CA, USA) and filled with sterile isotonic saline solution. The 8.5 French catheter is introduced via the femoral vein using the Seldinger technique and placed into the Inferior Vena Cava. The used ¼" silicon tube and the rotation speed limitation keeps the maximum flow at 300 ml/hour. The patients were cooled to 32 °C; this temperature was kept for 24 hours and controlled by a bladder or pulmonary catheter. Passive rewarming was instituted after removal of the cooling catheter. For this study 17 patients (8 female and 9 male) were cooled. The mean age was 62.1 years ranging from 37 to 86 years. Causes of cardiac arrest were: acute myocardial infarction (n=8), COPD with acute asphyxia (n=2), lung edema due to cardiac decompensation with acute asphyxia (n=1), septic shock (n=2), drug intoxication (n=1) and cardiac arrest without retrospectively specified cause. The body temperature was lowered down to 33°C in the first two hours and was maintained at this level for about 20 hours. Nine patients have showed no or moderate brain damage. Serious hypoxic brain damage resulted in 8 patients. Six patients died as a consequence of cardiac shock or multi organ dysfunction. No serious adverse effects like tachycardia or coagulation dysfunction were observed. At the beginning, the lactate level is mostly raised, and its normalization at a normal value around 2 mmol/l is a good sign of metabolic recovery, thus a trigger to stop the perfusion. When the catheter removed, the body will be re-warmed passively in about 4–to 5 hours.

We found this technique safe and effective. There are advantages especially for the cardiac instable patient. The combination of an endovascular-cooling catheter with an extracorporeal heat exchanger and a heater-cooler device offers sufficient cooling capacity and avoids volume overload especially in unstable cardiac patients.

§ 6. Conclusion

Hypothermia is one of the oldest medical treatments. With the development of cardiac surgery, two concepts have been developed. Induced clinical hypothermia has been designed and established as a concept to organ protection: Body temperature was reduced to the point where circulatory arrest could be tolerated for the time required to carry out an open heart procedure. Extracorporeal circulation has been developed as a competing concept: Blood is removed from the patient's circulation, delivered to an artificial oxygenator, and then returns to the patient while the patient's heart and lung are clamped. Combining both concepts was the way to the tremendous success story of open heart surgery. Today, mild therapeutic hypothermia is recommended by international guidelines for patients who were successfully resuscitated after out-of-hospital cardiac arrest.

References

1.   Hippocrate: On ancient medicine. In: Richard Kapferer (editor): Die Werke des Hippokrates. Die hippokratische Schriftensammlung in neuer deutscher Übersetzung. Hippokrates Verlag, Stuttgart 1936.

2.   Robert Boyle: New Experiments and Observations touching Cold (1665). In: Michael Hunter and Edward B. Davis (editors): The Pickering Masters. Part I: Vol. 1–7. Pickering & Chatto Publishers, London 1999.

3. Hughes McLeave: The risk takers. Frederick Muller, London 1962.

4. Rehn L. Ueber penetrierende Herzverletzungen und Herznaht. Arch Klin Chir 1897;55:315-29

5. Jean Dominique Larrey: Memoires de Chirurgie Militaire, et Campagnes. Vol. 1. Smith, Paris 1812.

6. Jean Julien Cesar Legallois: Expériences sur le principe de la vie, notamment sur celui des mouvemens du coeur, et sur le siège de ce principe. D'Hautel, Paris 1812.

7. Bigelow WG. Hypothermia. Its possible role in cardiac surgery. Ann Surg 1950;132:849-66

8. Bigelow WG. Application of hypothermia to cardiac surgery. Minn Med 1954;37:181-5

9. William Gordon Bigelow: Cold Hearts. The Story of Hypothermia and the Pacemaker in Heart Surgery. McClelland and Stewart Ltd., Toronto 1984.

10. Lewis FJ, Taufic M. Closure of atrial septal defects with the aid of hypothermia; experimental accomplishments and the report of one successful case. Surgery 1953;33:52-9

11. Gibbon JH. Application of a mechanical heart and lung apparatus to cardiac surgery. Minn Med 1954;37(3):171-80

12. Drew CE, Anderson IM. Profound Hypothermia in cardiac surgery. Report of three cases. Lancet 1959;273:748-50

13. Kalke BR, Castaneda A, Lilleihei CW. A clinical evaluation of the new Temptrol disposable blood oxygenator. Experience in 150 consecutive patients undergoing cardiopulmonary bypass. J Thorac Cardiovasc Surg 1969;57:679-87

14. Holtzer M, et al. Mild Therapeutic Hypothermia to improve the neurologic outcome after cardiac arrest. N Eng J Med 2002;346:549-56

15. Nolan JP, Morley PT, Hoek TL, Hickey RW. Therapeutic hypothermia after cardiac arrest. An advisory statement by the Advancement Life Support Task Force of the International Liaison Committee on Resuscitation. Resuscitation 2003;57:231–235

16. Nolan JP, Soar J, Boettiger B, et al. European Resuscitation Council Guidelines for Resuscitation 2010. Section 1: Executive Summary. Resuscitation 2010;81:1219–76

17. Zorn H, Janusch M, Silber R, Werdan K, Buerke M. Therapeutic Mild Hypothermia after Cardiac Arrest using a Modified Heart Lung Machine. Presented at the 7th European Conference on Perfusion Education and Training, September 15, 2007, Geneva, Switzerland

Cryo Storage and transport under the criterion of practical application

Rolf A. Sommer

Liquid Nitrogen cooling

In the cryo storage of biological material we must go out in principle from a temperature range starting from -150 °C. Thus, the basic technology substantially purchases storage by means of liquid nitrogen. This is from energetic view concerning the working reliability and the most common procedure.

Three procedures differ in detail during storage but these differ especially in contact with the medium.

First, the di rect storage in liquid nitrogen. This method has some major drawbacks. E.g. not controllable cross-contamination, toxic effect on the stored material and resonance passed directly to the object. As the liquid nitrogen can penetrate into the tissue, changes in temperature or removal of objects from the tank result in liquid nitrogen entering the tube and expansion when heated, which can cause an explosion or release of biologically hazardous material.

Another possibility is the storage in the vapor phase. In chief using this method, a contact of the object with the nitrogen is also possible, but only as nitrogen vapor. A major advantage is the appropriate design of the container with respect to the controllability of the temperature using a pressure valve. But this implies structural adjustment of the storage container according to the pressure vessel regulations.

Fig. 1 Container cooled by nitrogen vapour

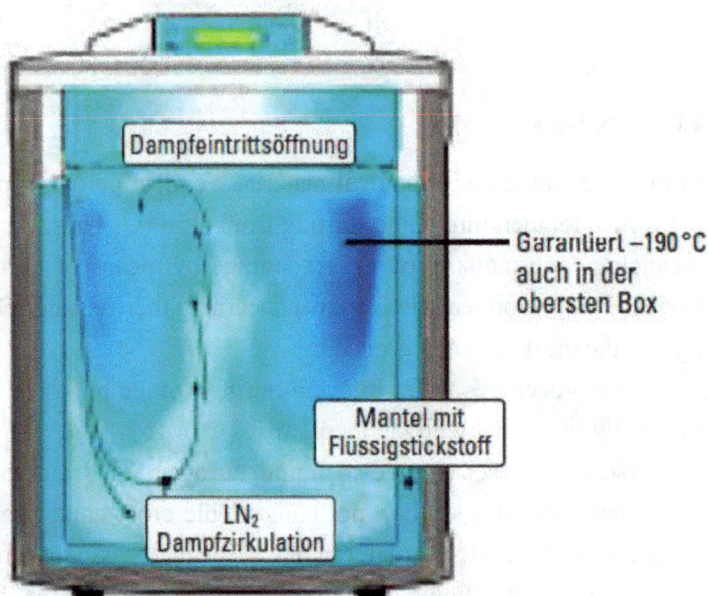

The outer cylinder filled with liquid nitrogen has an outlet opening into the internal cylinder containing the body, where N2 vapour circulates and can leave the tank via a pressure valve.

A third method which in practice is becoming increasingly important is the storage of the body outside of the vapor phase.

Fig. 2 Closed internal cylinder allows no contact of vapour to the body

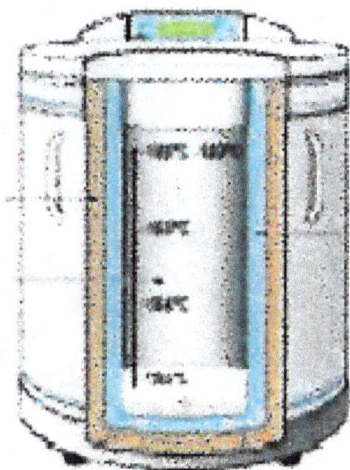

The outer most cylinder contains a vacuum phase. The liquid nitrogen occupies the next cylinder and the body is contained in the internal space.

This version in ideal perfection is only possible in very large containers, as we have seen a shift in temperature of 10° K over the entire storage area, which can add to tensions in the object. This can be prevented if you reduce the temperature to -150 ° C. Then you can record a constant temperature throughout the storage area.

The essential feature of all three variants, however, is the continuous monitoring and filling the tanks by staff.

Electrical generation of cryogenic cooling

In practice, even the type using a two-stage chillers would be cooling tissue to -165 ° C. One of the main advantages is the easy and service friendly handling. As a rule the systems runs smoothly at sustained power supply.

Fig. 3 Cooling box containing two stage chillers

The sizes are not given any Constructional constraints. Transport is also a given ability and use is subject to no rules. Interestingly this variant is hard to find as in the long-term storage, even if it provides advantages. Thus, long time securing of current is much easier compared to liquid nitrogen supply. In general, the insulating effect of electrical devices is higher than that of nitrogen storage tanks.

A predominant feature however, we want to set to transport temperature. At the moment a major shortcoming is still very poor supply of patients with cold during transport to the reservoir. We must also differentiate in what condition the patient is transported. Classically, the patient is transported on liquid ice. There already a start temperature of 0 ° C represents a problem.

In flight test over the distance Zurich - New York, a temperature change from +0 ° C to +14 ° C in 25 hours was detected.

Fig. 4 Dependence of temperature on time

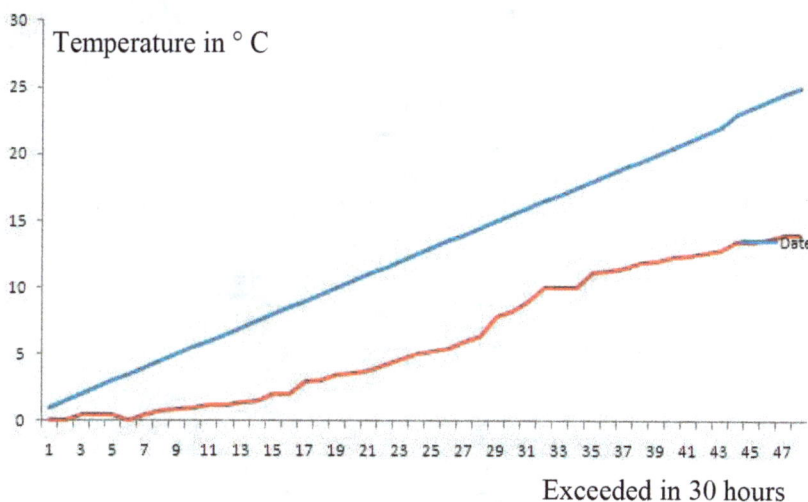

Exceeded in 30 hours

The graph shows curves of two data bases.

The transport of patients already stored in liquid nitrogen or dry ice with a previously received cryopreservation causes technical transportation problems. In line with the targets of the aviation authorities transport is limited by the evaporation of a flood of luggage space with nitrogen. The same holds true for transport in dry ice.

An alternative, transport refrigeration with Antifrogen

The transportation would however, be possible by Antifrogen, cooled using the ultra-dry ice to -68 ° C, the patient would be charged by NH 2 to -70 ° and at a starting temperature of -60 ° C following about 25 hours transportation in a super-insulated transport container a temperature rise up to -48 ° C would be recognizable.

Fig. 5 Antifrogen mediated temperature dependent on time

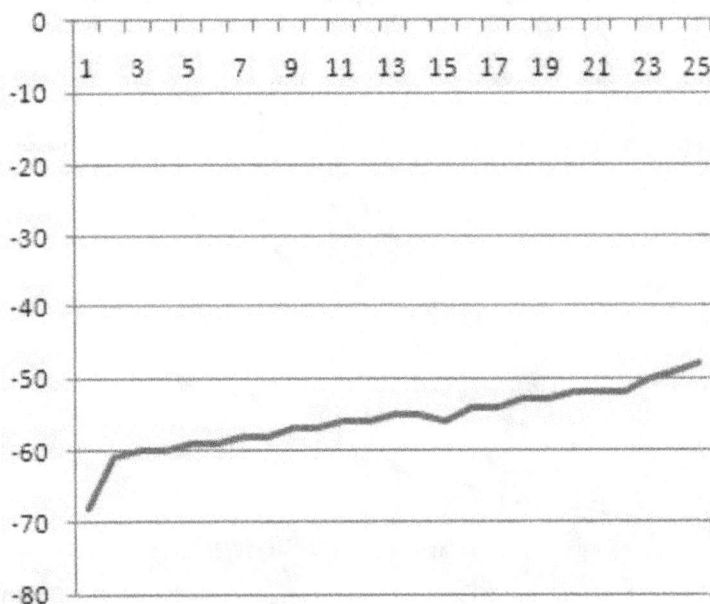

This is well below that of previous methods and guarantees a significant improvement. Because of the phase change we hold up to 18 hours temperatures below -55 ° C until this begin to climb upward. All this without changing the volume of the cooling medium in the main area. The Antifrogen used is absolutely anti-toxic and can be used repeatedly. The transport container can without expense be used corresponding to the regulations of the aviation authorities.

For the transportation of a "corpse" across the particular state boundary special conditions are placed on the coffin. This can be transported by land, by air or water. The body must lie in a hermetically sealed container, this is generally called the zinc coffin.

This tank must be filled with a material (sawdust, wood shavings, peat) suited to absorb solutions. For air transportation containers need a valve for pressure compensation, which cleans the air coming from the "corpse". A soldered zinc container is usually surrounded by a wooden coffin. The walls of the wooden coffin must be at least 20 millimeters thick. If this wooden coffin is from the outside soldered by a flat zinc layer or a material of different definition, 30 mm wood thickness are needed. Likewise, the number and spacing of the fittings have to be varied. This entire shipping container must then be placed into a neutral outer box so that the content can not be seen as a coffin.

The so-called zinc coffin is the actual container for transport. To the burial the dead is reburied in most cases. Zinc possesses anti bacterial properties, in connection with airtight sealing it prevents to fast a decay. The coffin is transported in the luggage compartment in principle, so that for no one of the surviving dependents access is possible. All these conditions are fulfilled. The time span is sufficient to perform regularly with the transport. This approach, however, requires a working infrastructure in Europe for patient care, including the Medical pre-requirements for cryopreservation.

Dry nitrogen tank

There is one more way of transporting with a so-called dry container. Initially, the container is filled with liquid nitrogen inside a ceramic material made of absorbent which absorbs the nitrogen. After about 20 minutes charging time, the excess nitrogen is discharged and the patient can be introduced into the container. It must be mandatory on it that he receives no contact with the storage medium. The container can now be sent.

The evaporation rate is in relation to the vacuum of your tank very low. Test has demonstrated a working time of up to 21 days. Such a small Taylor Wharton CX-100 tank has passed 20 days in the test. This variant allows the transport by ship and by different airlines at the transport request in Cargo transport by plane. It should be noted, however, that extreme pressure differences in air transport reduce the shipment duration of the container to 4 days. This variant will win in the next few years in the patient transport increasingly importance.

Thus we have here a small digest of the current situation in the appropriate patient positioning and transportation.

It would be important to opt for one of the conceptions, work it out consistently and fix it. Only then can a standard be created.

List of Contributors

Benjamin, P. Best, BSc, BBA, Degree in Pharmacy, President and CEO of Cryonics Institute Clinton Township, Michigan; Main Fields of Interest: Cryonics, Life Span Extension. Selected publications: Best, BP (2009). "Nuclear DNA damage as a direct cause of aging" (PDF). Rejuvenation Research 12 (3): 199–208; Benjamin P. Best (2008). "Scientific Justification for Cryonics Procedures" (PDF). Rejuvenation Research 11: 493–503; See. also at: www.benbest.com/

Chana De Wolf, M.S., Director, Researcher, Administrator of Neural Biosciences, Inc.; Research Areas: Cryonics, Neurophysiology, Laboratory Management; Portland, Oregon. Current results of research in preparation for publication.

Aschwin De Wolf, M.S., Director, Researcher, Administrator at Advanced Neural Biosciences, Inc.; Research Areas: Cryonics, Cerebral Ischemia, Depressed Metabolism; Research Design and Technical Writing; Portland Oregon. Current results of research in preparation for publication.

Gregory M. Fahy, Ph.D., Chief Scientific Officer and Vice President of 21st Century Medicine, Inc., Rancho Cucamonga, California; Cryobiologist; Research Areas: Cryobiology and Biogerontology. List of Publications s.: http://www.21cm.com/abstracts.jsp

Peter Gouras, M.D., Professor of Ophthalmology, Dept. of Ophthalmology; Research Areas: Age Related Degenerations in the Eye and the Effects of a Caloric Restricted Diet. Fields of interest e.g.: Macular Degeneration; Columbia University, New York, Selected publications: Escher P, Gouras P, Roduit R, Tiab L, Bolay S, Delarive T, Chen S, Tsai CC, Hayashi M, Zernant J, Merriam JE, Mermod N, Allikmets R, Munier FL, Schorderet DF. Mutations in NR2E3 can cause dominant or recessive retinal degenerations in the same family.. Hum Mutat. 2009 Mar;30(3):342-51; Gouras P, Ivert L, Mattison JA, Ingram DK, Neuringer M. Drusenoid maculopathy in rhesus monkeys: autofluorescence, lipofuscin and drusen pathogenesis. Graefes Arch Clin Exp Ophthalmol. 2008, 246 1403-11; Kong J, Kim SR, Binley K, Pata I, Doi K, Mannik J, Zernant-Rajang J, Kan O,

Iqball S, Naylor S, Sparrow JR, Gouras P, Allikmets R. Correction of the disease phenotype in the mouse model of Stargardt disease by lentiviral gene therapy. Gene Ther. 2008 Oct;15(19):1311-20. Epub 2008.

Torsten Nahm, Diploma in Mathematics, Biology and Computer Science (Studied), Specialization in Risk Modeling and Methodology at the Consulting and Auditing Company KPMG; Field of Interests: Future of Technology and its Social Implications; Co-founder of the German Transhumanist Association; Munich.

Klaus H. Sames, MD, Professor (German "Apl."); Anatomy and Experimental Gerontology, Institute of Anatomy University of Hamburg, Retired 2004; Field of Interest: Life Span Extension; Senden/ Iller (Germany). Selected publications: Sames K.: Function of proteoglycans in aging. Interdisciplinary Topics in Gerontology, Karger Verlag, Basel, 1994; Sames K.: Molekularbiologische Aspekte des Alterns. (no English, expertise for the report on the aged of the German federal government, authors translation of the title: Molecular biological aspects of aging); Sames K.: Altern, Fibrose und Reaktionsmechanismen des Bindegewebes. In: Ganten D., Ruckpaul K., Ruiz-Torres A. (eds.), Springer Verlag, Berlin 2004, pp. 402-28 (No English, author's translation: Springer Series on molecular medicine. Vol. Title: Molecular medical basis of age specific diseases. Title: Aging, fibrosis and reaction mechanisms of connective tissue).

Sebastian C. Sethe, Ph.D., Legal Advisor; Research in: Law, Regulation and Ethics of Regenerative and Personalised Medicine; London. Selected publications: Sethe SC: "Nanotechnology and life extension" In: F.Allhoff, P. Lin, J. Moor, J. Weckert (eds) Nanoethics: the social and ethical implications of nanotechnology. Wiley, 2007; pp. 353–365; Sethe SC: "The implications of "advanced therapies" regulation. Rejuvenation Res. 2010 Apr-Jun;13(2-3):327-8; Sethe SC, Miller J: "Gods with a limited budget: putting the utility back into utilitarian health politics" Interdisciplinary Science Reviews, Volume 30, 2005, pp. 273-278.

Rolf A. Sommer, Refrigeration Systems Specialist; Field of Interest: Cryogenic Cooling Devices; University Hospital Bern, Switzerland.

Jan Welke, StR, Established Graduate Secondary-School Teacher; Field of interest: Life span Extension and Cryonics; Leer (Germany).

Brian Wowk, Ph.D., Senior Scientist, Cryobiologist; Research Areas: Physics and Cryobiology, 21st Century Medicine, Inc., Rancho Cucamonga, California. For publication list s.: http://www.21cm.com/abstracts.jsp

Holger Zorn, Perfusionist, later Journalist; Research in Extra Corporal Technologies; Martin Luther University Hospital, Halle (Saale), Germany (till 2007). Selected publications: Scheubel RJ, Zorn H, Silber RE, Kuss O, Morawietz H, Holtz J, Simm A (2003). Age-dependent Depression in Circulating Endothelial Progenitor Cells in Patients Undergoing Coronary Artery Bypass Grafting. J Am Coll Cardiol 42:2073-80; Zorn H, Stiller M, Silber RE, Scheubel RJ. Extra Corporeal Systems With Surface Modifying Additives Do Not Improve Clinical Outcome After Coronary Artery Bypass Grafting (2005). J CARDIOVASC SURG 46(Suppl.1):159; Zorn H, Janusch M, Silber RE, Werdan K, Buerke M. Therapeutic Mild Hypothermia After Circulatory Arrest Using a Modified Heart Lung Machine (2007). Kardiotechnik 16(Suppl. 1):12.

ibidem-Verlag

Melchiorstr. 15

D-70439 Stuttgart

info@ibidem-verlag.de

www.ibidem-verlag.de
www.ibidem.eu
www.edition-noema.de
www.autorenbetreuung.de